A MATTER OF DEGREES:

THE POTENTIAL FOR CONTROLLING
THE GREENHOUSE EFFECT

Irving M. Mintzer

WORLD RESOURCES INSTITUTE
A Center for Policy Research

Research Report #5
April 1987

Kathleen Courrier
Publications Director

Myrene O'Connor
Marketing Manager

Hyacinth Billings
Production Supervisor

Gary Thompson
Cover Photo

Each World Resources Institute Report represents a timely, scientific treatment of a subject of public concern. WRI takes responsibility for choosing the study topics and guaranteeing its authors and researchers freedom of inquiry. It also solicits and responds to the guidance of advisory panels and expert reviewers. Unless otherwise stated, however, all the interpretation and findings set forth in WRI publications are those of the authors.

Contents

Acknowledgments

Many people have contributed in significant ways to this report. Gus Speth, Jessica Mathews, and Andy Maguire guided the analysis from its initial formulation to the final draft. Alan Miller and Rafe Pomerance strengthened the conceptual framework, sharpened the arguments, and provided crucial insights essential for resolving many thorny issues. Other members of the WRI staff including Peter Thacher, Jim Mackenzie, Mohamed El-Ashry, Mark Kosmo, Bob Repetto, Lauretta Burke, Lani Sinclair and Bob Kwartin provided critical reviews and comments.

The analysis profited broadly from reviews by a number of colleagues outside WRI. The author especially appreciates the comments provided by Dean Abrahamson, N. Dak Sze, Jae Edmonds, John Firor, Michael Gibbs, Peter Gleick, Gordon Goodman, John Harte, John Hoffman, Lars Kristoferson, Dan Lashoff, Michael MacCracken, Gordon MacDonald, V. Ramanathan, Joe Steed, Steve Seidel, C.C. Wallen, Wei-Chyung Wang, Bob Watson, John Wells, Bob Williams, Katy Wolf, George Woodwell, and Gary Yohe.

The clarity of the manuscript was vastly improved by editorial suggestions from Kathleen Courrier and Amber Leonard. The typing and production were made possible through the prodigious efforts of Dorothy Gillum, Cynthia Veney, Hyacinth Billings, and Allyn Massey. Any errors still present despite all this help are solely the responsibility of the author.

This work was made possible by the generous support of the Rockefeller Brothers Fund, the C.S. Mott Foundation, and the German Marshall Fund.

I.M.M.

Foreword

For the past several years the international scientific community has been issuing unusual warning signals. Earth's climate, they say, the climate that has sustained life throughout human history, is now seriously threatened by atmospheric pollution.

Perhaps the most notable warning came in October, 1985, as a conference sponsored by the International Council of Scientific Unions, the World Meteorological Organization, and the United Nations Environment Programme, in Villach, Austria, drew to a close. ''As a result of the increasing concentrations of greenhouse gases,'' the conference statement began, ''it is now believed that in the first half of the next century, a rise of global mean temperature could occur which is greater than any in man's history.''

More recently, in 1986 and 1987, the Environment and Public Works Committee of the U.S. Senate asked a dozen leading scientists from the United States and abroad to testify at extensive hearings. Wallace S. Broecker, a geochemist at Columbia University, spoke for many:

The inhabitants of planet earth are quietly conducting a gigantic environmental experiment. So vast and so sweeping will be the impacts of this experiment that, were it brought before any responsible council for approval, it would be firmly rejected as having potentially dangerous consequences. Yet, the experiment goes on with no significant interference from any jurisdiction or nation. The experiment in question is the release of carbon dioxide and other so-called greenhouse gases to the atmosphere.

Climate shapes the character of places and human activities as powerfully as any force. It allows corn to flourish in Iowa and causes crops to fail in sub-Saharan Africa. It makes Siberia harsh, Tahiti gentle, and lends distinctive character to every place on the globe.

Until recently, climate was taken for granted and assumed to be unchangeable. The evidence is clear, however, that through such activities as burning fossil fuels, leveling forests, and producing certain synthetic chemicals, people are releasing large quantities of ''greenhouse'' gases into the atmosphere. These gases absorb Earth's infrared radiation, preventing it from escaping into space. This process traps heat close to the surface and raises global temperatures.

Excess carbon dioxide (CO_2) is the main offender. Prior to the Industrial Revolution, the concentration of CO_2 in the atmosphere was about 280 parts per million. At this concentration, CO_2 (and water vapor) warmed Earth's surface by about 33°C and made Earth habitable. But, since then, especially since 1900 or so, the accelerating use of fossil fuels and vegetation loss over large areas of the planet have caused CO_2 in the atmosphere to increase by about 25 percent, to 346 parts per million.

But carbon dioxide build-up is not the only problem. Much of the new urgency about the greenhouse effect stems from the realization that other gases released through human activity—including chlorofluorocarbons (CFCs), methane, nitrous oxide, and others now contribute about as much to the greenhouse effect as CO_2 does.

According to one estimate, past emissions of greenhouse gases have *already* committed Earth to warm by 0.5°C to 1.5°C over the pre-industrial era, though only a fraction of this warming has been felt to date because of the inertia of the oceans. Several models project that if current trends in greenhouse gas build-up continue, we will have committed Earth to a warming of 1.5°C to 4.5°C by around 2030, the upper end of this range being the more probable.

It is vital to understand what these changes could mean. Not since the dawn of civilization some 8000 years ago has Earth been about 1°C warmer than today. To find conditions like those projected for the middle of next century, we must go back millions of years. In short, if the greenhouse effect turns out to be as great as predicted by today's climate models, and if current emission trends continue, our world will soon differ radically from anything in human experience.

While the regional impacts of a global warming are uncertain and difficult to predict, many of the anticipated changes are far-reaching and disturbing. Rainfall and monsoon patterns could shift dramatically, upsetting agricultural activities worldwide. In summer, the Great Plains of the United States, Central Europe, and parts of the Soviet Union could experience Dust Bowl conditions. Sea level could rise from one to four feet, flooding coasts and allowing salt water to intrude into water supplies. Ocean currents could shift, altering the climate of many areas and disrupting fisheries. The ranges of plant and animal species could change regionally, endangering protected areas and many species whose habitats are now few and confined. Record heatwaves and other weather anomalies could harm susceptible people, crops, and forests.

In this context, it is not surprising that the scientists at Villach took the important step of urging that the greenhouse issue be moved into the policy arena. ''[U]nderstanding of the greenhouse question is sufficiently developed,'' they concluded, ''that scientists and policy-makers should begin an active collaboration to explore the effectiveness of alternative policies and adjustments.''

WRI senior associate Irving Mintzer, who participated in the Villach meeting, has worked since 1985 to design a method to help scientists and policy-makers collaborate. *A Matter of Degrees* is the first fruit of that effort. Between scientific understanding and policy initiative lies the domain of policy research, and Mintzer's paper describes a promising tool for addressing many of the questions that must be answered before effective national and international action can be taken.

Mintzer has integrated many existing simulation models into one structure—the Model of Warming Commitment—that can be used to project future emissions of the six gases that contribute most to global warming and to estimate their ultimate warming effects. Individual components of the model can be separately manipulated to simulate possible policy initiatives and economic changes.

A Matter of Degrees describes the model and presents important results of analyses using it. These results provide grounds for great concern about current trends and guidance toward a less risky future. They suggest that strong measures can significantly diminish the build-up of greenhouse gases *if* implemented effectively and soon. They also suggest that even with such measures substantial climate change stemming from past and current activities may now be inevitable.

A comparison of two of Mintzer's scenarios brings out these conclusions. His ''base case'' scenario reflects conventional wisdom about population, economic, and energy growth from 1980 to 2075. No policies are introduced to slow the greenhouse gas build-up: no major effort is made to arrest tropical deforestation, to improve energy end-use efficiency, or to introduce non-fossil energy sources. Environmental concerns have little effect on energy policy. Similarly, no policy steps are taken to limit CFC emissions. In Mintzer's ''slow build-up'' scenario, economic growth is the same, but strong policies are adopted to introduce solar energy, improve energy efficiency, discourage the use of solid fuel, and arrest deforestation. CFC production is frozen by international agreement at the 1985 level, and other measures reduce CFC releases.

By 2030, the difference in global warming between these two scenarios is already significant in the model, with a warming commitment in the range of 1.6°C to 4.7°C in the base case scenario above pre-industrial temperatures and 1.1°C to 3.2°C in the slow build-up scenario. By 2075, the difference is profound—2.9°C to 8.6°C above pre-industrial temperatures in the former and 1.4°C to 4.2°C in the latter.

Using these results, it appears that, even with measures as far-reaching as Mintzer considers in the slow build-up scenario, the earth's surface temperature could be committed to an increase of 2°C or 3°C by about 2030—a major change in the planet's climate. Yet, the strong measures tested in the slow build-up scenario do succeed over the next century in preventing changes possibly two to three times this great, an extremely important result. Without such measures, current trends and policies appear to have mankind on a course that could lead to very large and potentially devastating climate changes in the next century.

Several responses to these findings are possible. Given the gaps in our knowledge and the inevitable distance between model and reality, one could simply beg to differ. Indeed, because the findings presented here are seriously disturbing, we must hope that future research and improvements in our understanding will lift this burden from societies in some measure. Time, of course, would tell whether the concerns now voiced by many scientists are justified. But can we afford to take that time and risk the planet's future? Mintzer's work, and that of others, suggests that, far from having the luxury of time, we are already late getting started and the costs of delaying a response are large.

Another possible response, considering the difficulty of measures needed to slow the build-up of greenhouse gases, is to conclude that not much can be done because of practical and political constraints. The greenhouse issue may present the ultimate environmental dilemma. Collective judgments of historic importance must be made—by decision or default—largely on the basis of scientific models that have severe limitations and that few can understand. To some, the trade-off will be whether to provide the energy needed for economic growth or to protect humanity from a seemingly distant and uncertain threat. Further, addressing that threat will require international cooperation on a scale seldom achieved save in war.

Yet, with the build-up of greenhouse gases proceeding apace, a great planetary experiment is under way, as Wallace Broecker correctly observed. Before the results are fully known, our children and future generations may have been irrevocably committed to an altered world—one that may be better in some respects but that also involves truly unprecedented risks. Not to despair requires an act of faith in people—faith that as these risks become widely appreciated, people will not sit idly by but will respond to protect deeply held values. Moreover, the alternative of action may appear more attractive as it is better understood. Virtually all the steps societies should take to slow greenhouse gas build-up have powerful justifications apart from preventing climate change. And to those who believe ''a little warming might not be so bad, on the whole,'' the appropriate response is that ''a little warming'' is no longer the issue. A big warming is.

A Matter of Degrees suggests that two kinds of action are justified: adaptive measures to prepare for climate change that seems inevitable and, even more important, preventive measures to forestall changes that we can still influence. Control of greenhouse gas releases can both buy precious time and, ultimately, maintain Earth's climate as close as possible to that of the past several thousand years. Given what is now known, major national and international initiatives—grounded in the best available science and policy analysis—should become a top priority of governments and citizens.

Clearly, a deeper appreciation of the risks of greenhouse gas build-up should spread to leaders of government and business and to the general public, investing the greenhouse issue with a sense of urgency not present today. If nations are to be spared a Hobson's choice between energy shortages and climate change, a priority commitment should be made here and abroad to energy efficiency, to solar and other new and renewable energy sources, and to economic incentives and other measures that discourage the use of high-carbon fuels. Steps should also be taken to halt the damaging deforestation now under way in the tropics and to regulate CFCs and other greenhouse gases.* *A Matter of Degrees* uses changes in all these directions to shape the slow build-up scenario.

The years immediately ahead should be a period of intense scientific research, policy exploration, and adoption of appropriate measures. Innovative international responses should be discussed. Global and regional energy futures should be explored with special emphasis on their effects on the greenhouse problem. Preventive and adaptive strategies appropriate to the world's regions should be mapped and means found to build U.S.-Soviet cooperation on this issue since these two countries together have 55 percent of the world's coal reserves.

In this period of intense interaction among scientists and policy-makers, scientists will have a special role to play. Without their leadership, the necessary public understanding will not be achieved. But the search for solutions must also involve leaders in government, business, energy, environment, and international and economic affairs. Cooperation will make the difference between choice and chance.

The World Resources Institute expresses its deep appreciation to the C.S. Mott Foundation, the Rockefeller Brothers Fund, and the German Marshall Fund of the United States for their support of WRI's work on energy and atmospheric issues.

James Gustave Speth
President
World Resources Institute

---

*See, e.g., *Tropical Forests: A Call For Action* and *The Sky Is The Limit*.

I. Introduction

Since the mid-19th century, scientists have recognized that carbon dioxide and certain other atmospheric gases allow incoming solar radiation to pass through the atmosphere but absorb and re-emit the low-energy radiation emanating from Earth's surface, warming the lower atmosphere (the troposphere) in the process.[1] Effluents from industrial and agricultural activities, particularly the combustion of fossil fuels, are increasing the concentrations of these gases in the atmosphere. The resulting global warming—the "greenhouse effect"—threatens to alter Earth's future climate in ways that scientists do not now completely understand.

How can the effects of private and public policy on the timing and extent of future global warming be evaluated? One new tool—described here—is a simulation model for projecting future emissions of the six greenhouse gases that contribute most to global warming and the resulting temperature increases. In this first exercise with this new Model of Warming Commitment, four scenarios (a base case and three "policy driven" scenarios) were constructed to illustrate how policy strategies implemented in the 1980s could affect the rate and magnitude of future global warming. Projections based on these scenarios indicate that *policies implemented soon and continued over the next several decades could significantly affect the rate and extent of global warming due to greenhouse gas build-up.*

Several analysts have recently suggested that using energy more efficiently and altering the mix of commercial fuels burned worldwide would keep emissions of carbon dioxide (CO_2) and other trace gases significantly lower than current trends imply.[2] Similarly, national and international efforts to slow tropical deforestation may affect the biotic contribution to CO_2 emissions (and perhaps to methane emissions as well) by changing the balance between biomass burning and forest growth. In addition, policies to limit the risk of stratospheric ozone depletion can significantly change the magnitude and composition of chlorofluorocarbon emissions to the atmosphere.

How much could such policies affect the timing and extent of global warming? To find out, several existing models of energy use and trace gas emissions have been combined into the WRI Model of Warming

Commitment to link scenarios of global economic growth with estimates of future emissions of the most important greenhouse gases.[3] Whereas most previous research in this field has focussed on the links between future levels of economic activity and future emissions of CO_2, the Model of Warming Commitment links future emissions of chlorofluorocarbons (CFCs) and nitrous oxide (N_2O) to the scenarios of economic growth, energy use, and CO_2 emissions. Because so much uncertainty remains about the sources and sinks of methane (CH_4) and tropospheric ozone (O_3), future

---

Whereas most previous research in this field has focussed on the links between future levels of economic activity and future emissions of CO_2, the Model of Warming Commitment links future emissions of chlorofluorocarbons (CFCs) and nitrous oxide to the scenarios of economic growth, energy use, and CO_2 emissions.

---

emissions of these gases are not yet linked in the model to estimates of future economic activity. Instead, simple exponential growth projections based on recent trends are used for these two gases. The Model of Warming Commitment is modular so alternative methods can be used to project future energy use and economic activity (or to link these activities to future emissions).

Numerous recent studies have explored trace gas build-up and global warming. Seidel and Keyes (1983) in their pathbreaking work, "Can We Delay a Greenhouse Warming?," used the same energy-economic model applied here and concluded that energy policies could alter the rate of future CO_2 emissions but could not significantly affect the date at which the planet was committed to a warming of 2°C. Rose and his colleagues (1983) used the same model to illustrate a range of energy scenarios with CO_2 emissions both significantly higher and substantially

1

lower than those described by Seidel and Keyes. Ramanathan *et al.* (1985) attempted to quantify the direct warming effects of future emissions of more than twenty radiatively active trace gases. If current emissions trends continue, Ramanathan concluded, the non-CO_2 trace gases could amplify the CO_2-induced warming by a factor of 1.5 to 3.0. MacDonald (1986), Cicerone and Dickinson (1986), and the World Meteorological Organization (WMO) (1986a) have also used emissions scenarios to estimate the combined warming effects of CO_2 and other radiatively active trace gases. The WMO (1986b) report suggests that, if current emissions trends continue, Earth could be committed as early as the 2030s to a global warming equivalent to that which would result from doubling the pre-industrial concentration of CO_2—a benchmark level used here and by other scientists to indicate significant change. Except for the early work of Seidel and Keyes, none of these studies considers the effects of policy on future emissions rates.[4]

Unlike these earlier studies, *A Matter of Degrees* combines consideration of future emissions of CO_2, N_2O, and the two principal CFCs with projections of future energy use and explores the effects of broadly defined global policies on the rate and extent of future commitments to global warming.

The four scenarios of greenhouse gas emissions investigated here are presented as illustrations of the range of possible future outcomes, not as predictions or forecasts. Each scenario incorporates both broad policy strategies (combinations of policies that change trends in deforestation, technological innovation and energy use) and such narrowly defined policy measures as taxes on particular fuels or limits on certain uses of CFCs.

When key assumptions in the base case are changed, the model can simulate the impacts of broad policy strategies. The parameters varied in the present study include:
1. the rate of improvement in the efficiency with which energy is used (efficiency of energy end-use);
2. the price and availability of unconventional sources of energy including synfuels and solar technologies;
3. the rate of tropical deforestation and land-use conversion; and
4. the impact of changes in income levels and energy prices on future energy demand.

Five specific policy measures are tested in these four scenarios. They include: 1) consumption taxes on commercial energy use proportional to the carbon content of the fuel consumed; 2) environmental surcharges incorporated into the production price of energy supplies; 3) limits on the production of CFC-11 and CFC-12; 4) limits on specific uses of CFC-11 and CFC-12; and 5) mandatory controls on CFC losses in manufacturing and disposal operations.

The effects of these policies were translated into trajectories of future emissions. The model then simulated the processes of atmospheric retention and removal for each chemical species to obtain estimates of the future concentration of each gas. Finally, the combined warming effect of these increased concentrations was estimated for each scenario.

The four scenarios presented in this study illustrate the range of likely greenhouse futures and establish the credibility of the simulation model. The next step will be to refine a realistic menu of policy options and test the effects in a variety of combinations.

II. The Greenhouse Problem

Many economically important human activities emit gaseous pollutants into the air. Some of these emissions, including carbon dioxide and certain other gases with absorption bands in the range of 8 to 13 microns, alter the heating rates in the atmosphere, causing the lower atmosphere to warm and (in the case of CO_2) the stratosphere to cool. This phenomenon is commonly called "the Greenhouse Effect" because, like the glass roof of a greenhouse, these tropospheric gases temporarily trap heat that would otherwise rapidly escape into space.

Many different gases can induce tropospheric warming via the greenhouse effect. Some are highly stable and linger in the atmosphere for decades or even a century or more. (See Table 1.) The most important of these gases are water vapor, CO_2, methane (CH_4), nitrous oxide (N_2O), tropospheric ozone (O_3), and the chlorofluorocarbons (especially $CFCl_3$, commonly known as CFC-11, and CF_2Cl_2, commonly known as CFC-12). The atmospheric concentrations of each of these gases has been increasing since the beginning of the industrial era. (See Figure 1.) Other atmospheric pollutants including certain aerosols and particulates may also affect future climate regimes. Most scientists agree that these factors may have significant effects on regional climates; none will have as large an effect on global climate as the build-up of greenhouse gases. Historically, CO_2 and water vapor have contributed most to the greenhouse effect. Indeed, the increasing concentration of CO_2 has added more to global warming since the Industrial Revolution than have changes in the concentration of any other trace gas.

Some of these other greenhouse gases, however, absorb infrared radiation up to 10,000 times more efficiently than CO_2 does on a per-molecule basis. In the last several decades, the balance between the effects of CO_2 build-up and the effects of other greenhouse gases has changed. Recent analyses indicate that together these other trace gases now contribute about as much annually to global warming as CO_2 does. In the future, greenhouse gases other than CO_2 are likely to contribute more than half of the total commitment to global warming.

Over the last ten million years, the naturally occurring concentration of CO_2 has fluctuated substantially. Throughout this period, CO_2 and water vapor in the atmosphere have warmed the planet's surface. Together, clouds, water vapor, and pre-industrial concentrations of 275-285 ppmv of CO_2 warmed Earth's surface by approximately 33° centigrade—from an estimated average temperature of −18° (in the absence of CO_2) to approximately +15°C.[5] This background

---

Ironically, one of the great forces of evolution—this same Greenhouse Effect—now threatens to disrupt human societies and natural ecosystems.

---

greenhouse effect figured centrally in the evolution of Earth's present climate, elevating the average surface temperature to a level between that of ice and steam. Save for this background effect, Earth would be a comparatively cold and lifeless planet.

Ironically, this same greenhouse effect now threatens to disrupt human societies and natural ecosystems. During the last century, anthropogenic emissions of carbon dioxide and other greenhouse gases have altered the atmosphere, which had been stable for thousands of years. Fossil fuel combustion (along with other industrial and agricultural activities) has caused the atmospheric concentration of carbon dioxide to increase approximately 25 percent since about 1860.[6] The combined atmospheric build-up of carbon dioxide and the other greenhouse gases since 1860 are believed to have already committed Earth's surface to warm approximately 0.5° to 1.5°C above the average global temperature of the pre-industrial period.[7] For perspective, a change in average global temperature of only 1°C separates the current climate regime of North America and Europe from that of the Little Ice Age of the 13th to 17th Centuries.

Even small changes in average global temperatures can have large effects. At a 1985 meeting on the Greenhouse Effect sponsored by the WMO, United

---

A change in average global temperature of only 1°C separates the current climate regime of North America and Europe from that of the Little Ice Age of the 13th to 17th Centuries.

Nations Environment Programme (UNEP), and The International Council of Scientific Unions (ICSU), scientists from twenty-nine countries declared:

> *Many important economic and social decisions are being made today on major irrigation, hydro-power and other water projects; on drought and agricultural land use, on structural designs and coastal engineering projects; and on energy planning, all based on assumptions about climate a number of decades into the future. Most such decisions assume that past climatic data, without modification, are a reliable guide to the future. This is no longer a good assumption since the increases of greenhouse gases are expected to cause a significant warming of the global climate. It is a matter of urgency to refine estimates of future climate conditions to improve these decisions.*

Continuing emissions of CO_2 and other greenhouse gases will commit the atmosphere to significant future warming. Most atmospheric scientists agree that when the atmospheric concentration of CO_2 reaches approximately 550 ppmv, Earth will be committed to an average warming of 1.5°-4.5°C above pre-industrial temperatures.[8] Recent experiments with the most advanced general circulation models of the atmosphere suggest that the planet's temperature sensitivity to doubled CO_2 is likely to be in the top half of this range,

---

A global warming of even 1.5°C could alter Earth's climate to an extent outside the range observed in the last 10,000 years.

---

between 3.0° and 4.5° C.[9] If current emission growth trends for all trace gases continue, the *combined* effects of the six most important greenhouse gases could, possibly as early as 2030, commit the globe to warm as much as a doubling of the pre-industrial concentration of CO_2.[10] A global warming of even 1.5°C over pre-industrial levels could alter Earth's climate to an extent outside the range observed in the last 10,000 years.

Table 1

Name	Formula	Residence Time, years	1985 Concentration	Annual Growth Rate
Carbon Dioxide	CO_2	2–3	345 ppmv	0.5%
Nitrous Oxide	N_2O	150	301 ppbv	0.25%
Methane	CH_4	11	1650 ppbv	1.0%
CFC-11	$CFCl_3$	75	0.20 ppbv	7.0%
CFC-12	CF_2Cl_2	111	0.32 ppbv	7.0%

Table 1   Important Greenhouse Gases

Figure 1a. Estimated Atmospheric Concentration of CO_2, (Parts per Million by Volume)

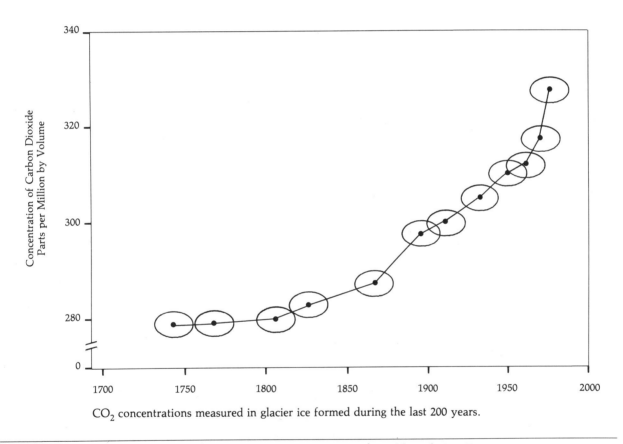

CO_2 concentrations measured in glacier ice formed during the last 200 years.

Source: Neftel, et al., "Evidence from Polar Ice Cores for the Increase in Atmospheric CO_2 in the Last Two Centuries," *Nature*, volume 315, May 2, 1985.

Figure 1b. Estimated Atmospheric Concentration of N_2O (Parts per Billion by Volume)

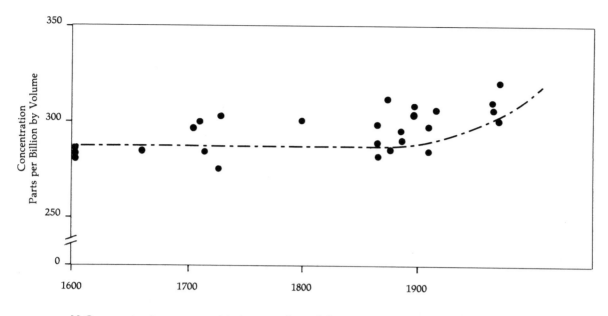

N₂O concentrations measured in ice cores formed during the last 400 years.

Source: Pearman et al., "Evidence of Changing Concentrations of Atmospheric CO_2, N_2O, and CH_4 from Air Bubbles in Antarctic Ice," *Nature*, volume 320, 1986.

Figure 1c. Estimated Atmospheric Concentration of CH_4 (Parts per Million by Volume)

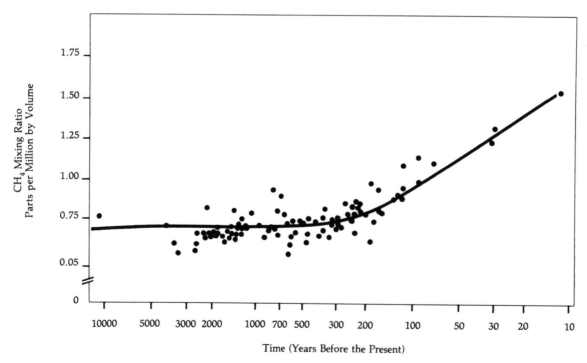

CH₄ mixing ratios measured in air trapped in ice cores formed during the last 10,000 years

Source: Bolle et al., "Other Greenhouse Gases and Aerosols," in Bolin, et al., *The Greenhouse Effect, Climatic Change, and Ecosystems*, John Wiley and Sons, Chichester and New York, 1986

III. The Model of Warming Commitment

The Model of Warming Commitment simulates the effects of various policy strategies on the build-up of greenhouse gases in the atmosphere in three stages. First, future production and emissions of the most important greenhouse gases are projected. Second, these emissions are translated into estimates of future atmospheric concentration, reflecting the natural removal rates for these gases. Third, the combined radiative effects of the build-up of these greenhouse gases are evaluated and the commitment to future warming estimated.

The model comprises several smaller, specialized sub-models linked together to generate and analyze internally consistent scenarios. *(See Figure 2.)* Among the primary inputs to the model are assumptions about future population levels, global and regional energy resources, energy and materials use, and certain policy-dependent assumptions about taxes, income and price elasticities of demand, and other economic factors. These assumptions drive the sub-models of economic activity that projects the regional fuel mixes and computes estimates of CO_2, N_2O and CFC production and use.

In the cases of CO_2 and N_2O, each year's production is assumed to be released to the atmosphere immediately. In the case of the CFCs, regional production estimates are fed into another sub-model that allocates production to specific end-uses so that "prompt" and "banked" emissions can be separated.[11] Because the processes controlling tropospheric increases of methane and ozone are less well-understood, future methane concentrations are estimated using a simple exponential growth model while tropospheric ozone concentration is assumed to increase at a linear rate reaching a maximum of 15% above 1980 levels in 2040.

In the second stage of the analysis, atmospheric retention models are used to simulate the chemical and biological processes by which CO_2, N_2O, and the CFCs are removed from the atmosphere. The resulting projections of future trace gas concentrations are estimated on an annual basis.

In the third stage, the model estimates the radiative effects of these projected increases. Each gas is treated separately using the 1980 concentration as the benchmark. These warming effects are added to the past warming commitment due to emissions from 1860 to 1980. The total warming effect (without atmospheric feedbacks) is then computed as the sum of the effects of the individual gases.[12]

Various physical feedback processes at work in the atmosphere will amplify or diminish the warming commitment from this ensemble of greenhouse gases. For example, as the air warms, the water vapor content of the atmosphere, the extent of sea ice, the average extent and height of cloud cover, the Earth's surface albedo (or reflectivity) and other characteristics change. Because of continuing uncertainty about the net effects of these feedback processes, this study estimates a range rather than a precise value for the final warming effect.

A. Carbon Dioxide

Future atmospheric concentrations of carbon dioxide will be determined by CO_2 emissions from the combustion of fossil fuels, by emissions of CO_2 from biotic systems, and by the workings of the global carbon cycle. The Model of Warming Commitment estimates future emissions of CO_2 from fossil fuels by simulating the evolution of the market for all commercial energy sources. Future emissions of CO_2 from the biota are estimated separately. The global carbon cycle and its effects on the retention of CO_2 in the atmosphere are simulated through the use of an airborne fraction model.

The energy-economic sub-model uses an economic partial equilibrium analysis (simulating markets for energy but not for other goods) to balance long-term projections of energy supply and demand. Developed at the Institute for Energy Analysis (IEA) by Jae

Figure 2. Schematic Structure of the Model of Warming Commitment

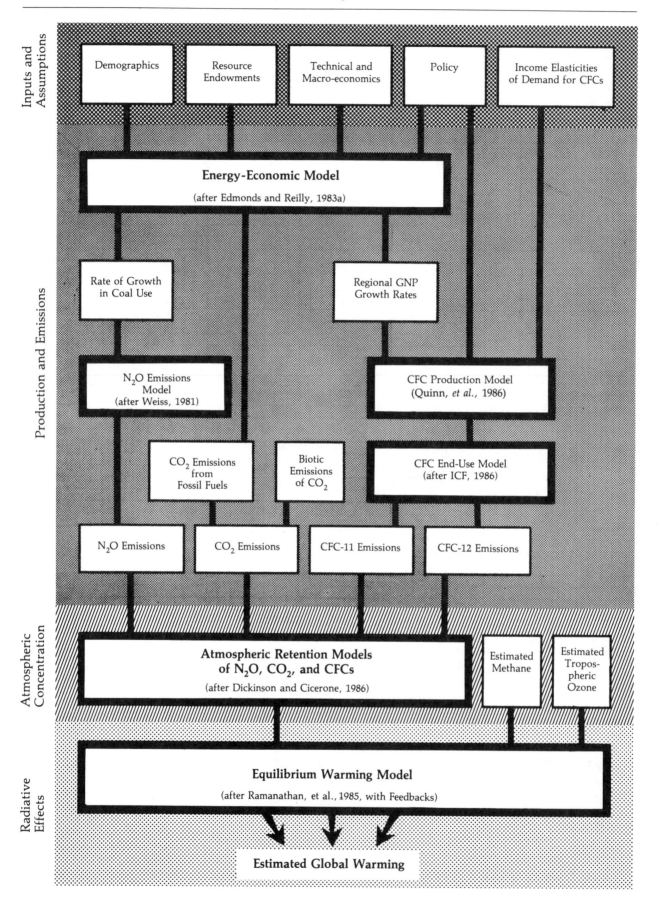

Edmonds and John Reilly for the U.S. Department of Energy, this widely used model facilitates the study of alternative long-term energy strategies, their effects on global economic activity, and their implications for future CO_2 emissions from commercial fuel use.[13]

1. Inputs and Assumptions about Energy Use

The IEA energy-economic sub-model divides the world into nine geopolitical regions. *(See Figure 3.)* One important structural assumption made in the IEA energy model is that all energy forms except electricity can be traded between regions. (This simplifying assumption ignores the fact that electricity is currently traded between regions—for example, between the United States and Canada.)

Demographic factors are the principal determinants of GNP growth in the energy-economic model. Intermediate estimates of "base GNP" are derived primarily from projections of future regional population levels, labor force participation rates, and rates of change in labor productivity.[14] Assumptions about these rates are held constant in the four scenarios used here, so GNP levels change little from scenario to scenario. In short, the model evaluates how much policies adopted in the next few decades can affect the energy use rate and the extent of CO_2 emissions under comparable conditions of GNP growth.

Regional supplies of nine primary and four secondary energy forms are calculated in the energy-economic model for each forecast period. The model divides the nine sources of primary energy into three categories and estimates the cost of supplying each category.

The three categories are resource-constrained non-renewable fuels (conventional oil and gas), resource-constrained renewable technologies (hydroelectricity and biomass), and unconstrained supply technologies (coal, nuclear energy, synthetic fuels derived from coal and shale oil resources, and solar energy).

The magnitude of regional resource endowments in each of five grades of fuel and the minimum extraction costs of each grade are used to estimate the prices of such conventional fuels as coal, oil, natural gas, and nuclear electricity. These resource endowments remain constant in all four scenarios. The supply of hydroelectricity is not determined by price and thus the total estimated global resource is phased in gradually over time. The estimated costs of producing biomass are based on assumptions about the present and future costs of these resources, the size of the resource base, and the time required to reach the final (least-cost) production level. The cost and availability of hydro and biomass resources are the same in all four scenarios. In each forecast period, the production costs of unconstrained or "backstop" technologies (including solar electric systems and synthetic fuels) change over time. The trajectories follow logistic curves (determined by the assumed initial and ultimate supply prices and the years required to reach minimum cost of supply). It

is assumed that these technologies will be introduced when their cost falls below that of competing conventional substitutes.

Regional levels of GNP and population determine the "base demand" for energy services. This base demand is then adjusted to estimate "final demand" by applying different scenario-dependent assumptions about income and price elasticities, interfuel substitution, and the increasing efficiency of energy supply and use.

For each of the world's industrialized regions, the model uses one coefficient to represent the rate of change in the efficiency of energy use in each of three economic sectors.[15] For the developing countries, where only one end-use sector is represented in the model, a single coefficient represents the average rate of improvement over the forecast period for all of a region's end-uses. A second set of coefficients represents the rate of improvement in the efficiency of energy supply.[16] For each region, this rate is specified for conventional oil and gas, coal, nuclear energy, and unconventional oil.

2. Outputs of the Energy-Economic Model

In addition to regional estimates of GNP in each forecast period, the model outputs include estimates of future fuel prices, the global fuel mix, and the amount of CO_2 released from fossil fuel combustion. (The final fuel prices estimated by the model are the average world price for each type of primary fuel.) The model also specifies the quantities of primary and secondary energy in each region during each period. The amount of CO_2 released annually from fossil fuel combustion is computed by multiplying the quantities of each fuel consumed by the estimated CO_2-intensity of the fuel. *(See Table A-1 and Figure A-1 in the Appendix.)*

The biotic emissions of CO_2, estimated exogenously, follow time-dependent trajectories designed to illustrate alternative global policies on deforestation and land-use conversion. Because the role of the biota in the global carbon cycle is not fully understood, the range of estimates for net biotic emissions after 2000 is large, differing by a factor of 150 between the High Emissions and Slow Build-up scenarios in 2075. *(See Figures A-1 and A-2.)* Total annual emissions of CO_2 in each scenario are the sum of the fossil fuel and biotic emissions.

In the model-generated scenarios, specific policy measures (for instance, consumption taxes on particular fuels) can be represented explicitly and can be varied on a regional basis. In contrast, other model parameters are proxies for the effects of a combination of policy measures. For example, a policy to speed the introduction of a specific technology would likely involve an array of measures—among them, direct support for research and development, preferential tax treatment, and subsidized credit programs. In the model, the effects of such policy combinations are expressed in terms of the time it takes for the new

Figure 3. Geopolitical Regions in the Energy-Economic Model

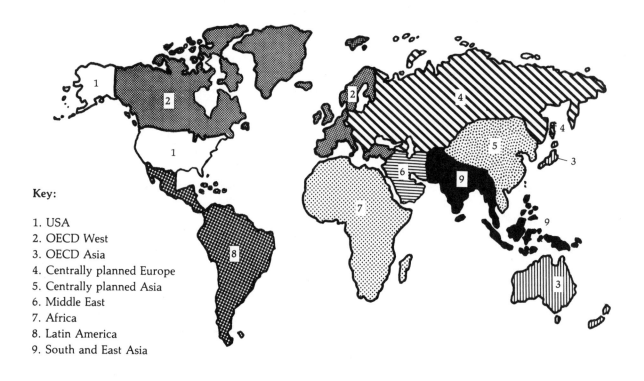

Key:

1. USA
2. OECD West
3. OECD Asia
4. Centrally planned Europe
5. Centrally planned Asia
6. Middle East
7. Africa
8. Latin America
9. South and East Asia

Source: Edmonds, J., and J. Reilly, 1983a. "A Long-Term Global Energy-Economic Model of Carbon Dioxide Release from Fossil Fuel Use," in *Energy Economics,* volume 5, number 2, 1983.

technology to reach its minimum production cost. Thus, this single proxy variable represents the effects of policy measures that share the same objective but take different forms in different countries.

3. Future CO_2 Concentrations

Total annual CO_2 emissions are converted to future atmospheric concentrations by a simple airborne fraction model.[17] In this model, the concentration of carbon dioxide in year 't' is a function of three parameters—the concentration of CO_2 in year 't-1', the total emissions of CO_2 in year 't', and the fraction of emissions from year 't' that remain airborne in the atmosphere. The relationship among these factors is as follows:

$$C(t) = C(t-1) + AirbornFrac*[ConvFactor* Emit(t-1)]$$

where: $C(t)$ = the concentration of CO_2 in year t, i.e., given in parts per million by volume, (ppmv)

AirbornFrac = the fraction of CO_2 emitted in year t-1 which is still airborne in year t, i.e., 0.58

Emit(t) = total emissions of CO_2 in year t, in gigatons

ConvFactor = a factor to convert gigatons to ppmv, i.e., 0.471

B. Other Greenhouse Gases

1. Nitrous Oxide

Nitrous oxide is an extremely stable compound that resides in the atmosphere for 150 years on average.[18] In the troposphere, it is an important greenhouse gas. When it rises into the stratosphere, it helps destroy Earth's ozone shield. The historic growth rate in N_2O concentration correlates closely with growth in the use of fossil fuels, especially coal and fuel oil.[19] In this study, future N_2O emissions in each scenario were assumed to increase at a rate equal to the annual rate of growth in coal combustion.[20] Recent research suggests that deforestation may also be a significant source of future N_2O emissions.[21] Due to lack of detailed data, these sources are not modelled in this analysis. The emissions and concentration of N_2O in the base year (1980) were derived from World Meteorological Organization estimates.

The sub-model used to convert future emissions of nitrous oxide to estimates of atmospheric concentration is based on a model developed by Craig and colleagues[22] and later modified by Weiss[23] to fit historic concentration and emissions data. The estimated steady-state concentration of N_2O in the unperturbed atmosphere was taken from Weiss.[24] In this model, the future concentration of N_2O is computed as a function of the previous year's concentration, current annual emissions of N_2O, the assumed atmospheric lifetime of N_2O, and the pre-industrial concentration of N_2O.

The expression used to calculate the concentration of N_2O in year 't' is:

$$C(t) = C(t\text{-}1) + [EMIT(t)/CONVFACTOR] - [((C(t\text{-}1) - SSCONC) * 1 \text{ YEAR})/LIFETIME]$$

where:

$C(t)$ = the concentration of N_2O in year t, ppbv

$SSCONC$ = the steady-state concentration of N_2O in the atmosphere prior to significant anthropogenic perturbations, ppbv

$CONVFACTOR$ = a factor to convert concentrations given in ppbv to the equivalent quantity of N_2O given in metric tons, assumed equal to approximately 7.9 million tons per ppbv

$EMIT(t)$ = emissions of N_2O in year t, tons

$LIFETIME$ = atmospheric residence time of N_2O, assumed to be approximately 150 years.

2. Chlorofluorocarbons

Although at least two dozen types of chlorofluoro-carbons are currently being produced and emitted to the atmosphere, CFC-11 and CFC-12 are the dominant species, and along with CFC-113, are likely to dominate future CFC production for at least several decades unless a global regulatory strategy is adopted.

The Model of Warming Commitment was used in this study to test the effects of two types of policies—production limits and use controls—on the projected future atmospheric concentrations of CFC-11 and CFC-12. Future production of these two compounds was estimated using an approach developed by the RAND Corporation.[25] The RAND method estimates future regional production for aerosol and non-aerosol applications of both CFC-11 and CFC-12 in each of the regions analyzed by the energy-economic model. (See Appendix Table A-2.)

In each region, for each scenario the CFC demand is projected for three periods of market development starting in 1990, based on the population and GNP estimates produced by the energy economic model. These three periods—early growth period, emerging market, and mature market—simulate the historical pattern of development of the CFC markets in the United States through varying assumptions of the income elasticity of demand. (See Table 2.) These variables are combined with estimates of population and regional GNP to determine future production levels for aerosol and non-aerosol applications in each region. The future income elasticities of demand for CFCs in each region outside the United States are assumed to be a function of the historical income elasticities of demand in the United States and vary among the scenarios.

Unlike CO_2 and N_2O, not all the CFCs produced in a year are promptly released into the atmosphere. Much of the CFC embodied in, say, refrigerators, air conditioners, and certain kinds of closed-cell foams remains trapped until the devices are finally scrapped years later. In other applications (among them, aerosol propellants and some open-cell foams), CFCs are released into the atmosphere during manufacture or within a year of their production.

To account for these differences in patterns of end-use and for losses in manufacturing, a model developed by ICF Incorporated has been used to convert estimates of future production to projections of future emissions of CFC-11 and CFC-12.[26] For four major categories of end-use—aerosols, foams, hermetically-sealed, and non-hermetically-sealed refrigerators—manufacturing losses, disposal losses, prompt emissions, and banked emissions in each year are estimated.

The atmospheric retention model used to convert estimates of future emissions to projections of future CFC concentrations was derived from Cicerone and Dickinson.[27] In this model, annual concentrations are a function of the atmospheric residence time of the particular CFC, the previous year's concentration, and emissions in the previous year. The relationship between these factors in the model is expressed below:

$$C(t) = C(t\text{-}1) + [EMIT(t) - (MASS(t\text{-}1)/LIFETIME)] * CONVFACTOR$$

where:

$C(t)$ = the concentration of CFC-x in year t, ppbv

$MASS(t\text{-}1)$ = atmospheric mass of CFC-x in year t-1, tons

$LIFETIME$ = the atmospheric residence time of CFC-x

$CONVFACTOR$ = a factor to convert emissions in tons to ppbv

$EMIT(t)$ = estimated emissions of CFC-x in year t, tons

Table 2. Historical Development of the U.S. CFC Market

Application	Time Period	Comments	Estimated Income Elasticity of Demand
CFC-11 Aerosol	1946–1952	Early development period	2.33
	1953–1975	Market for personal care products develops	2.96
CFC-11 Non-Aerosol	1935–1950	Early development period	1.83
	1951–1959	Market for refrigerators develops	3.30
	1960–1982	Market for foam products develops	4.39
CFC-12 Aerosol	1946–1952	Early development period	2.51
	1953–1975	Market for personal care products develops	2.84
CFC-12 Non-Aerosol	1935–1950	Early development period	2.62
	1951–1957	Market for refrigerators develops	3.02
	1958–1982	Market for mobile A/C develops	3.17

Source: Quinn et al., ''Projected Use, Emissions, and Banks of Potential Ozone-Depleting Substances,'' RAND Corporation Report No. N-2282-EPA, January 1986.

3. Methane

Future atmospheric concentrations of methane (CH_4) will be strongly affected by the magnitude of anthropogenic and biotic sources and by the processes of atmospheric removal (e.g. through reaction with hydroxyl radicals, OH). The availability of the atmospheric sink depends on competition with other atmospheric constituents (especially carbon monoxide, CO) for access to the existing stock of highly reactive OH radicals. Reaction with CO reduces the number of OH radicals and, in effect, extends methane's atmospheric lifetime.

Although some sources of CH_4 may be linked to patterns of energy use, others are believed to be independent. Since all the chemical and biological processes that affect methane sources and sinks are not well understood today, the rates of future emissions and removal are not mathematically simulated in the model. Instead, a simple exponential growth function is used.

These rates of future methane build-up in the WRI scenarios reflect the assumption that some sources of methane emissions (e.g., enteric fermentation in ruminant animals and anaerobic digestion in rice paddies, swamps, and bogs) remain constant in all scenarios, while others (e.g., leaks from natural gas pipelines) vary with patterns of future energy use. Another assumption is that factors affecting the atmospheric sinks, including carbon monoxide emissions from automobiles and other vehicles, could be affected by policy choices. Growth rates were selected to be consistent with the policy strategy applied to control emissions of other trace gases in each scenario.

4. Tropospheric Ozone

The processes that control the rate of change in concentration of tropospheric ozone are also complex and incompletely understood. Ozone concentration varies significantly by time of day, among geographic locations, with latitude, and with altitude. However, since no existing ozone-concentration model captures these effects, the concentration of tropospheric ozone was simply assumed to increase linearly, reaching a maximum in 2040. This approach ignores the effects on ozone concentration of interactions between CO_2, N_2O, CH_4, and CFCs and may thus understate the build-up.[28]

B. Temperature Effects of a Greenhouse Gas Build-up

The build-up of trace gases will alter the atmosphere's radiation balance by trapping in the troposphere long-wave radiation emitted from the Earth's surface and concurrently cooling the stratosphere.[29]

How much the lower atmosphere warms and Earth's average surface temperature changes depends primarily on greenhouse gas build-up, but other processes, including natural variations in solar output and the volcanic injection of aerosols into the stratosphere also play a role.[30] Injecting aerosols into the lower atmosphere will probably alter only local climate regimes.

Any warming from a greenhouse gas build-up will not register immediately as a change in surface temperature. The oceans' large thermal mass will cause a lag in the warming effect. Nonetheless, the build-up of greenhouse gases will cause an eventual or "equilibrium" warming, perhaps several decades after atmospheric concentrations first increase. The Model of Warming Commitment estimates this eventual warming effect using an approach developed by Ramanathan et al. (1985).

The Ramanathan model is calibrated to a 2°C-rise in temperature for an atmosphere with a CO_2 concentration of about 550 ppmv. To simulate feedback effects upon warming, the temperature increase is scaled by factors of 0.75—2.25.[32] For instance, for a build-up equivalent to that from doubled CO_2, the point estimate is scaled from 2°C to 1.5°—4.5°C.

As a check on these estimates of warming commitment, an alternative computation based on the work of Lacis and his colleagues[33] was applied to the 2075 trace gas concentrations in each scenario. Appendix Table A-3 compares the resulting range of warming commitment in 2075 for each scenario to the estimates produced using the Ramanathan approach.[34] The Lacis equation yields somewhat lower estimates for the high emissions case and higher estimates for all other cases. However, the differences between the two approaches do not significantly alter the conclusions of this analysis.

The Equilibrium Warming Model

The Model of Warming Commitment converts future concentrations of greenhouse gases to their warming effect by multiplying the change in concentration by the unit radiative effects of each gas. These calculations are based on a one-dimensional model developed by Ramanathan and his colleagues.[31] Analyzing the absorption and emission spectra of more than a dozen radiatively active trace gases, Ramanathan et al. estimate the direct warming effects due to changes in the concentration of each gas. (See Figure 4.) Ramanathan's results were refined in WRI's study to reflect the differing potency of greenhouse gases. Thus, warming due to increases in CO_2 concentration is scaled logarithmically; the effects of build-up in methane and nitrous oxide are scaled to the difference in the square root of the concentration in the future and the "reference" atmosphere. For tropospheric ozone and CFCs, the effects are scaled upward linearly from the values estimated by Ramanathan et al. The equations used to scale the temperature effects are as follows:

$$T_{CO_2} = -0.677 + 3.019 * \ln[CO_2(t)/CO_2(o)]$$
$$T_{N_2O} = 0.057 * [N_2O(t)^{0.5} - N_2O(o)^{0.5}]$$
$$T_{CH_4} = 0.019 * [CH_4(t)^{0.5} - CH_4(o)^{0.5}]$$
$$T_{O_3} = 0.7 * [(O_3(t) - O_3(o))/15]$$
$$T_{CFC-11} = 0.14 * [CFC11(t) - CFC11(o)]$$
$$T_{CFC-12} = 0.16 * [CFC12(t) - CFC12(o)]$$

where:

$CO_2(t)$ = CO_2 concentration in year t
$CO_2(o)$ = CO_2 concentration in 1880
$N_2O(t)$ = N_2O concentration in year 't'
$N_2O(o)$ = N_2O concentration in 1980
$CH_4(t)$ = CH_4 concentration in year 't'
$CH_4(o)$ = CH_4 concentration in 1980
$O_3(t)$ = O_3 concentration in year 't'
$O_3(o)$ = O_3 concentration in 1980
$CFC(t)$ = CFC concentration in year 't'
$CFC(o)$ = CFC concentration in 1980
(concentration of CO_2 in ppm, others in ppb)

The total radiative forcing (Ts) is estimated to be the sum of these effects:

$$T_S = T_{CO_2} + T_{N_2O} + T_{CH_4} + T_{O_3} + T_{CFC11} + T_{CFC12}$$

Figure 4. Atmospheric Sensitivity to Uniform Increases in Concentration

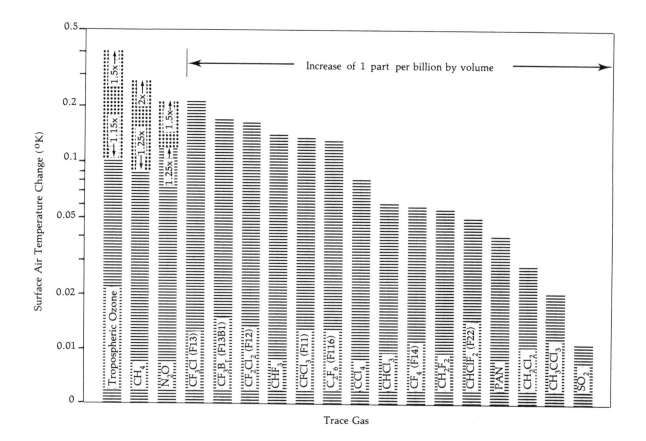

Computed surface air temperature change due to a 0 to 1 parts per billion by volume increase in trace gas concentrations. Tropospheric O_3, CH_4, and N_2O increases are also shown for comparison. For these three gases, the factor of increase is indicated in the figure.

Source: Ramanathan et al., "Trace Gas Trends and Their Potential Role in Climate Change," *Journal of Geophysical Research,* volume 90, number D3, June 1985

IV. Scenarios of Future Energy Use and CO_2 Emissions

The potential effects of policies on the future atmospheric concentration of atmospheric CO_2 can be illustrated by four scenarios that include both fossil fuel use and biotic emissions of CO_2. Integral to each scenario is a global policy strategy that could alter aggregate global demand for primary energy and the relative prices of commercial fuels. *(See Table 3.)* These relative price changes affect the mix and quantity of commercial fuels supplied in each region and place the global energy system on four very different paths of development.

The four illustrative scenarios reflect different levels of effort toward slowing a greenhouse warming. The Base Case Scenario embodies the conventional wisdom about a wide range of factors affecting world economic growth and commercial energy use over the next 90 years. In this scenario, the characteristics of today's energy sector change only incrementally, and no policies are introduced specifically to reduce the rate of greenhouse gas build-up. In the High Emissions Scenario, global energy use increases rapidly and potential environmental impacts due to global warming or other factors are largely ignored. The Modest Policies Scenario features a few measures to slow the rate of greenhouse gas build-up. The Slow Build-up Scenario represents a determined global effort to minimize the rate of greenhouse gas emissions and to limit some of the other environmental risks of energy use.

The level of energy use varies significantly in the WRI scenarios. *(See Figure 5.)* Primary energy use in the Base Case is approximately 520 EJ per year in 2025. In the other three scenarios, it ranges from approximately 250 EJ per year in 2025 in the Slow Build-up Scenario to 710 EJ per year in the High Emissions Scenario. In 2075, the Base Case demand for primary energy is 940 EJ per year, in the High Emissions Scenario approximately 1600, and in the Slow Build-up Scenario approximately 260 EJ per year. The four scenarios embody different energy price trajectories and different rates of growth in per capita energy use. *(See Tables A-5, A-6 and A-7 in the Appendix.)*

The differences in energy prices and per capita energy use among the four scenarios reflect differences in input assumptions and key model parameters. *(See Table 4.)* The magnitude and distribution of regional resource endowments are held constant, but the rates of technological progress in various activities and for competing energy supply systems differ by scenario. For example, the rates of improvement in energy supply efficiency—defined as the annual percentage decrease in the cost of energy supply and given in

---

None of the scenarios examined here will eliminate the risk of global warming due to increasing atmospheric concentrations of carbon dioxide or other greenhouse gases.

---

constant dollars per gigajoule—vary by factors of two to five. This cost reduction reflects improvements in the technology of energy supply (e.g., improved oil-drilling technology). Similarly, the rate of improvement in end-use efficiency varies by a factor of 7.5 between the High Emissions and the Slow Build-up Scenarios.

These illustrative scenarios are not a forecast of how the global energy sector might or should evolve. But, because they encompass the range of energy estimates found in most earlier reports, the future will probably evolve within the range defined by these scenarios.

None of the scenarios examined here will eliminate the risk of global warming due to increasing atmospheric concentrations of carbon dioxide or other greenhouse gases. All of the WRI scenarios assume continued release of fossil CO_2, varying in 2075 from about 16 gigatons (Gt) per year in the Base Case up to 30 Gt per year in the High Emissions Scenario and down to 2.0 Gt per year in the Slow Build-up Scenario, compared to the current release rate of about 5 Gt of carbon per year. *(See Figures*

Table 3.   Energy Policies in the WRI Scenarios

	Related Energy Model Parameter Value
Base Case Scenario	
• ''Business-As-Usual,'' the inertial model of growth and change in the world energy industry	
• No policies to slow carbon dioxide emissions	
• Minimal stimulus to improve end-use efficiency	(Rate of change = 0.8% per year)
• Modest stimulus for synfuels development	(Final Price = $3.15-$4.25 per GJ in 2005)
• Minimal stimulus for development of solar energy systems	(Final Price = $16.50 per GJ in 2025)
• No policy to limit tropical deforestation or to encourage reforestation	
• Minimal environmental costs included in price of energy	($0.30 per GJ for coal; $1.00 per GJ for synfuels)
High Emissions Scenario	
• Accelerated growth in energy use is encouraged	
• No policies to slow carbon dioxide emissions	
• No stimulus to improve end-use efficiency	(Rate of change = 0.2% per year)
• Modest stimulus for increased use of coal	(Rate of improvement = 0.75% per year)
• Strong stimulus for synfuels development	(Final Price = $2.75-$3.50 per GJ in 1995)
• No stimulus for development of solar energy systems	(Final Price = $20 per GJ in 2040)
• Rapid deforestation and conversion of marginal lands to agriculture	
• Token environmental costs included in price of energy	($0.15 per GJ for coal; $0.50 per GJ for synfuels)
Modest Policies Scenario	
• Strong stimulus for improved end-use efficiency	(Rate of change = 1.0% per year)
• Modest stimulus for solar energy	(Final price = $15.00 per GJ in 2025)
• Substantial efforts at tropical reforestation and ecosystem protection; more intensive rather than extensive agriculture encouraged	
• Substantial environmental costs imposed on energy prices to discourage solid fuel use and encourage fuel-switching	($0.60 per GJ for coal; $1.50 per GJ for synfuels)
Slow Build-up Scenario	
• Strong emphasis placed on improving energy efficiency	(Rate of improvement = 1.5% per year)
• Rapid introduction of solar energy encouraged	(Final Price = $12.00 per GJ in 2000)
• Major global commitment to reforestation and ecosystem protection	
• High environmental costs imposed on energy prices to discourage solid fuel use and encourage fuel-switching	($1.20 per GJ for coal; $3.00 per GJ for synfuels)

A-1 and A-2 in the Appendix.) Total CO_2 emissions in the four scenarios are mapped in Figure 6.

A. The Base Case Scenario

In this ''conventional wisdom'' future, the price of coal increases by a factor of 1.7 (in constant dollars) from 1975 to 2075, while oil and gas prices increase 2.7 and 4.9 times, respectively. *(See Appendix Table A-5.)* Per capita energy use increases by 0.7 percent per year in the industrialized countries and by 1.1 percent annually in the developing countries. *(See Appendix Table A-6.)* Total primary energy use increases by about 1.3 percent per year from 1975 to 2075, growing from approximately 250 EJ per year to 940 EJ per year. *(See Appendix Table A-7.)*

Electricity use increases in proportion to other

carriers, growing from 13 percent of secondary energy use in 1975 to 37 percent in 2075. During the same period, liquid fuels decline from 45 percent of secondary supplies in 1975 to 32 percent one hundred years later. Environmental concerns do not significantly limit growth in energy use.

Table 5 compares the global primary energy use in 2025 for the Base Case Scenario with a range of published estimates for that year. (The various estimates are not exactly comparable since different assumptions about population, economic growth, and future patterns of energy supply and consumption are used.[35]) Energy use in the WRI Base Case Scenario is lower than in earlier estimates primarily because the average global rate of growth in gross national product (GNP) per capita is now lower.

In addition to population and per capita GNP, energy use in the model is affected most by changes in the

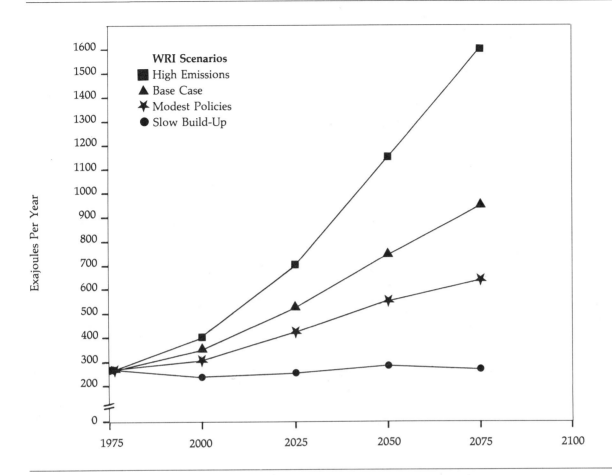

efficiency of energy supply and the prices of fuels. The price of each fuel estimated by the model is the sum of extraction costs, environmental costs, transportation costs, and taxes on energy use. Modest improvements in the efficiency of energy end-uses and in the efficiency of energy supply are assumed in the Base Case. *(See Table 4.)* On the demand side, energy intensities decline by 0.8 percent per year.

In this scenario, improvements in coal technology reduce the price of coal-derived energy by 0.2 percent per year. Estimated environmental costs for coal use (approximately $0.30 per GJ or $1.80 per barrel of oil equivalent, BOE)[36] are incorporated into the consumer price. The development of shale oil and coal-derived synthetic fuels is encouraged. The environmental costs of energy supply are kept to a minimum for unconventional oil (approximately $1 per GJ or $6 per BOE), and minimum production costs of $4.25 per GJ of synthetic oil and $3.15 per GJ of synthetic gas are achieved in the year 2005.

Environmental costs of $7.50 per GJ are imposed on nuclear electricity supply systems, and only limited investments are made to improve this technology. The

efficiency of nuclear supply improves at the same rate as that of coal. Development of solar technology is not encouraged, and thus the final costs of energy from this source approximately equal four times the cost of synthetic oil ($16.50 per GJ for solar vs. $4.25 per GJ for synthetic oil) after fifty years. As a result of these assumptions, natural gas use grows by roughly 1.7 percent per year to 2000 then levels off. The use of solid fuels increases substantially (by approximately 2.1 percent per year) from 1975 to 2075. Nuclear energy supply grows from 7 EJ per year in 1975 to about 40 EJ in 2075. Significant amounts of solar energy first penetrate the global economy in 2025 (with use estimated at 13 EJ per year); commercial use of solar energy grows to 38 EJ in 2075. Global use of hydro resources ultimately grows to about 120 GJ in 2075, approximately three times the estimated level of nuclear power use in that year.[37]

Compared to the historic energy mix in 1975, the energy mix in 2025 for the Base Case Scenario features the market share of oil and gas declining in favor of solid fuels (i.e., coal plus biomass) and hydro power. *(See Figures 7 and 8.)* Beyond that, the Base Case

Table 4   Input Parameters for Energy Scenarios

Input Parameter		High Emissions	Base Case	Modest Policies	Slow Buildup
Demographic factors					
Population in 2075, billions	(1)	10.5	10.5	10.5	10.5
Average Annual Increase in Labor Productivity					
Industrial Countries	(2)	1.2–2.3%	1.2–2.3%	1.2–2.3%	1.2–2.3%
Developing Countries		1.6–1.9%	1.6–1.9%	1.6–1.9%	1.6–1.9%
Technical and Macroeconomic Factors					
Average Annual Increase in the Efficiency of Energy Use	(3)	0.2%	0.8%	1.0%	1.5%
Average Annual Increase in Energy Supply Efficiency					
Oil	(4)	0.60%	0.30%	0.30%	0.60%
Gas		0.30%	0.30%	0.40%	1.50%
Coal		0.75%	0.20%	0.20%	0.20%
Nuclear		0.20%	0.20%	0.30%	0.40%
Unconventional Oil		0.75%	0.30%	0.30%	0.15%
Synthetic Oil	(5)				
Final Cost, 1975 US$		$3.50	$4.25	$5.00	$7.00
Years to Final Cost		20	30	45	60
Synthetic Gas	(6)				
Final Cost, 1975 US$		$2.75	$3.15	$4.00	$5.50
Years to Final Cost		20	30	45	60
Solar Energy	(7)				
Final Cost, 1975 US$		$20.00	$16.50	$15.00	$12.00
Years to Final Cost		65	50	50	25
Income Elasticity of Energy Demand	(8)				
OECD		1.00	1.00	0.90	0.80
East Bloc		1.00	1.00	1.00	0.80
Developing Countries		1.40	1.40	1.25	1.10
Aggregate Energy Price Elasticity	(9)	−1.1	−0.8	−0.7	−0.7
Policy Factors:					
Environmental Cost of Energy Supply, '75$/GJ					
Oil		$0.00	$0.00	$0.00	$0.75
Gas		$0.00	$0.00	$0.00	$0.55
Coal		$0.15	$0.30	$0.60	$1.20
Unconventional Oil		$0.50	$1.00	$1.50	$3.00
Nuclear		$7.50	$7.50	$7.50	$10.00
Consumption Tax on Energy End Uses					
Oil		0%	0%	0%	7%
Gas		0%	0%	0%	5%
Coal		0%	0%	0%	10%
Electricity		0%	0%	0%	8%

Notes to Table 4

1. The population projections used in this study are World Bank estimates derived from Vu (1985).
2. Nordhaus and Yohe (1986) surveyed twelve recent studies containing estimates of annual increases in labor productivity. They derive estimates for global average rates of increase of 1.05–2.95% per year from 1980–2000; 0.75–2.85% per year for 2000–2025; and 0.30–1.70 for 2025–2050.
3. Edmonds *et al.* (1985) suggest a range for the energy efficiency parameter of 0.0–3.0% per year. In a recent survey of oil use in the residential sector of seven OECD countries, Schipper and Ketoff (1985) observed rates of decline in energy intensity of between 2.2 and 5.1% per year sustained over a nine to eleven year period after 1972. The rate of increase in energy efficiency for these end uses in the U.S. was approximately 4.4% per year.
4. Edmonds *et al.* (1985) suggest a range of estimates for the rate at which technological changes reduce the price of commercial energy supplies. The range suggested for coal is −0.5 to 2.0% per year; for unconventional oil, −0.5 to 2.5% per year; and for nuclear, 0.0 to 1.5% per year.
5. Edmonds *et al.* (1985) suggest a range of estimates for the future cost of synfuels between $3.50 and $20.00 per GJ in 1975$. The estimates used in this study are at the low end of this range.
6. Edmonds *et al.* (1985) suggest a range of estimates for the future cost of synfuels between $3.50 and $20.00 per GJ in 1975$. The estimates used in this study are at the low end of this range.
7. Edmonds *et al.* (1985) suggest a range of estimates for the future costs of solar energy of between $6 and $200 per GJ. A review of estimates prepared by the U.S. Energy Information Administration, SRI International, and the U.S. National Academy of Sciences suggests a range of $5.90 to $112 per GJ. The values used in this study are at the low end of this range.
8. Edmonds *et al.* (1985) suggest a range of estimates for the future value of the income elasticity of demand for energy of 0.2–2.2. In this study we apply values from the middle of this range, 0.8–1.4.
9. Edmonds *et al.* suggest a range of estimates for the future aggregate price elasticity of demand for energy of between −0.05 and −1.3. Nordhaus and Yohe (1983) indicate a range of −0.7 to −1.2 and use a "best guess" value of −0.7. In this study, we apply a range of values from −0.7 to −1.1.

Table 5   Comparison of Base Case Scenario With Other Mid-Range Projections of Future Energy Demand, 2025

Source	Population in billions	Global Primary Energy Demand EJ Per Year	GROWTH RATES	
			Pop	GNP Per Capita
Perry & Landsberg (1978)	9.3	1230	1.7%	1.8%
Edmonds & Reilly (1982)	7.4	910	1.2%	1.7%
Nordhaus & Yohe (1983)	7.8	760	1.4%	1.9%
Seidel & Keyes (1983)	7.4	710	1.2%	1.7%
Edmonds & Reilly (1984)	7.4	673	1.2%	1.7%
Base Case Scenario (This Study)	8.2	520	1.5%	1.4%

Sources: J. Edmonds and J. Reilly, 1982.
        J. Edmonds and J. Reilly, 1984.
        W. Nordhaus and G. Yohe, 1983.
        J. Perry and H. Landsberg, 1978.
        S. Seidel and D. Keyes, 1983.

Figure 6. Total Emissions of CO_2 in the WRI Scenarios (Gigatons of Carbon per Year)

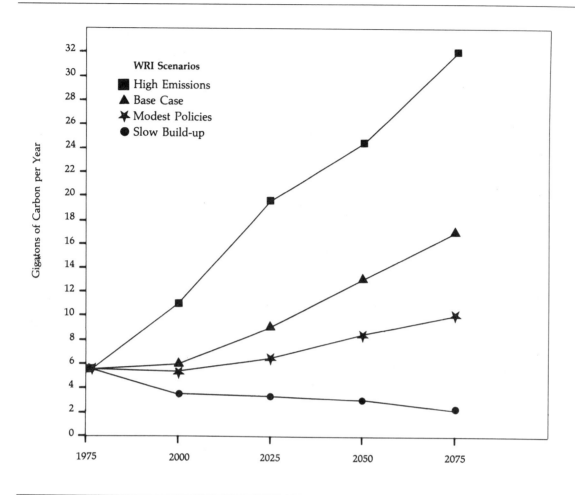

Scenario leads to an energy economy in 2025 and 2075 that differs little from today's. The trend toward electrification continues and the aggregate share of fossil fuels in the energy mix declines. Fossil fuels represent over 70 percent of primary energy supplies in 2025 (and in 2075) compared to over 90 percent in 1975. The combined share of solar, hydro, and biomass increases from about 7 percent in 1975 to approximately 20 percent in 2025 and 2075 while the market share of liquid fuels declines.

Biotic releases remain constant at 1.0 Gt per year. In the Base Case, CO_2 emissions from commercial fuel use grow from 4.5 Gt of carbon in 1975 to approximately 8 Gt in 2025 and nearly 16 Gt in 2075. The total annual emissions of CO_2 to the atmosphere include releases from both the biota and fuel combustion.

B. The High Emissions Scenario

This scenario resembles a projection made by the World Energy Congress.[38] In it, national policy-makers ignore the environmental risks of energy use, including

the Greenhouse Effect. *(See Table 3 and Table A-8.)* Table 4 compares some key input assumptions in the High Emissions case to the three other WRI scenarios.

The price of coal doubles between 1975 and 2075, while the prices of oil and gas increase by factors of 2.5 and 5.1, respectively. *(See Appendix Table A-5.)* Total primary energy use increases by 2.1 percent per year from 1975 to 2025 and 1.7 percent annually thereafter, reaching 710 EJ per year in 2025 and approximately 1600 EJ per year in 2075. *(See Figure 9.)* Per capita energy use increases in this scenario by 1.2 percent per year in the industrialized countries and by 1.6 percent per year in the developing countries. *(See Appendix Table A-6.)*

In this scenario, electricity use grows to 26 percent of secondary energy demand in 2025 and to 36 percent in 2075. The share of liquid fuels in total secondary energy demand declines from 45 percent in 1975 to 34 percent in 2025 and to 30 percent in 2075.

Primary energy use and total CO_2 emissions are greater in 2025 in the High Emissions Scenario than in the Base Case. Because energy intensities decline on average by only 0.2 percent per year (compared to 0.8 percent per year in the Base Case), more primary

Figure 7. Global Primary Energy Supply, 1975

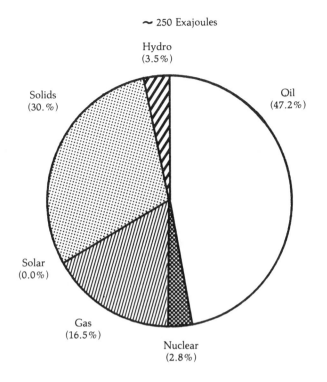

~ 250 Exajoules

Figure 9. High Emissions Scenario

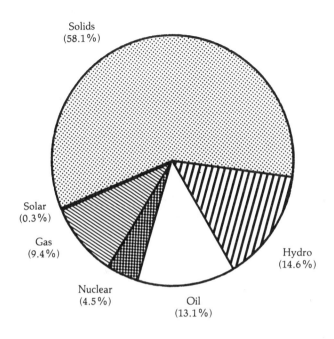

Primary Energy Supply, 2025, ~ 710 exajoules

Figure 8. Base Case Scenario

Primary Energy Supply, 2025, ~ 520 exajoules

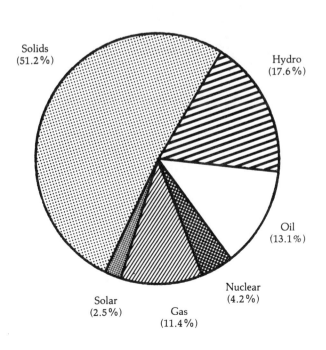

Figure 10. Modest Policies Scenario

Primary Energy Supply, 2025, ~ 420 exajoules

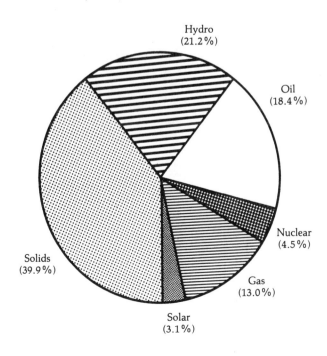

21

Figure 11. Slow Build-Up Scenario

Primary Energy Supply, 2025, ~ 250 exajoules

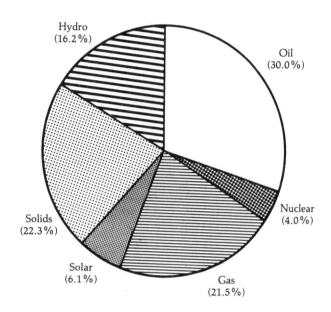

Hydro (16.2%)
Oil (30.0%)
Nuclear (4.0%)
Solids (22.3%)
Solar (6.1%)
Gas (21.5%)

energy is required to support an equivalent level of economic growth. On the supply side, a broad array of policies is implemented to speed the introduction of coal and other solid fuels. Together, these policies increase the efficiency (and reduce the cost) of coal supply by an average rate of 0.75 percent per year, compared to 0.2 percent annually in the Base Case. *(See Table 4.)* The efficiency of unconventional oil supply increases at 0.75 percent per year, compared to 0.3 percent annually in the Base Case. Policies to encourage the introduction of synthetic fuels help bring the production costs of these technologies down to their minimum in 20 years, compared to 30 years in the Base Case.

The same minimal environmental penalties are imposed for using nuclear energy in the High Emissions as in the Base Case Scenario. Environmental surcharges for coal and shale oil are reduced to $0.15 and $0.50 per GJ, respectively, from $0.30 and $1.00 per GJ in the Base Case.

The efficiency of nuclear supply grows at the same rate as in the Base Case. The introduction of solar technology is not encouraged, so the production costs of this technology do not fall to their minimum for 65 years, compared to 50 years in the Base Case Scenario.

As a consequence of these assumptions, coal use increases more than five fold by 2025, growing by 3.4 percent per year and increasing by 2075 to more than fifteen times the 1975 level. Nuclear energy use also grows in the High Emissions Scenario (by approximately 2.4 percent per year).[39] Fission-electric

systems which accounted for less than 3 percent of primary energy supply in 1975, account for 4.5 percent of the total by 2025; by 2075 they provide approximately 5 percent of primary energy and deliver more than ten times as much primary energy as they did in 1975. In this simulation, solar energy first penetrates the market significantly in 2025. Solar systems contribute less than 0.3 percent of primary energy supplies in 2025; the solar share grows to approximately 3 percent of total primary energy in 2075, compared to 4.6 percent in the Base Case Scenario. *(See Figure 9 and Appendix Table A-7.)*

Comparing Figure 9 to Figure 7 reveals a decline in the market share of oil and gas in 2025 similar to that seen in the Base Case. The share of primary energy represented by solids, especially coal, is greater in the High Emissions Case than in the Base Case, while the contribution of hydroelectric power is somewhat less. *(See Figure 8.)*

In this scenario, CO_2 emissions increase rapidly because of dependence on coal and shale. Coal supplies nearly 75 percent of the primary energy in 2075 for the High Emissions Scenario (compared to 66 percent in the Base Case in 2075), up from a historic level of 30 percent in 1975. In 2025 and in 2075, the renewable technologies (solar, hydro, and biomass), supply about 15 percent of primary energy, compared to over 20 percent in the Base Case Scenario.

The High Emissions Scenario assumes that tropical forests and other unmanaged or lightly managed ecosystems are rapidly converted to agriculture and other purposes. In addition, there is a temperature-driven feedback assumed only in this scenario. Biotic releases of CO_2 are assumed to rise as higher ambient temperatures increase respiration of soil bacteria. No effort is made to increase the efficiency of combustion for non-commercial biomass either to reduce fuelwood demand or to minimize emissions of carbon monoxide (CO), and CO_2. As a consequence, the biotic contribution to atmospheric CO_2 emissions is assumed to increase from 1.0 Gt of carbon per year in 1975 to 7.5 Gt of carbon in 2025 and then to decline to 1.5 Gt in 2075 as forest supplies are exhausted.

C. The Modest Policies Scenario

In this scenario, global concern about the possible impacts of a greenhouse warming leads to modest policy changes. *(See Table 3.)* The price of coal increases by a factor of 1.5 between 1975 and 2075. During the same period, the prices of oil and gas increase by factors of 3.0 and 4.3, respectively. *(See Table A-5.)* Per capita energy increases by 0.27 percent per year in the industrialized countries and 0.67 percent per year in the developing countries from 1975 to 2075. *(See Table A-6.)* Total primary energy use increases by approximately 1.0 percent per year from 1975 to 2025 (compared to 1.5 percent per year in the Base Case), reaching 420 EJ per year by 2025. From 2025 to 2075, primary energy use grows to 630 EJ per year (equivalent to an annual rate of about 0.8 percent per year), compared to 1.2 percent per year in the Base Case. *(See Table A-7.)*

Electricity use grows from 13 percent of total secondary energy supply in 1975 to 26 percent in 2025 and 37 percent in 2075—roughly the same shares as in the Base Case. The market share of liquid fuels declines from 45 percent in 1975 to 26 percent one hundred years later, compared to 30 percent in 2075 in the Base Case. Figure 5 illustrates the level of primary energy use in the Modest Policies Scenario compared to the other WRI scenarios. Table A-9 compares the Modest Policy Scenario with several published "low energy" futures.

CO_2 emissions from fossil fuel combustion after 2000 are 27 percent to 42 percent lower in the Modest Policies Scenario than they are in the Base Case, mainly because the use of solid fuels is reduced. Emissions from commercial energy use are estimated to be 5.6 Gt of carbon per year in 2025 and 9.1 Gt in 2075, compared to 7.7 Gt and 15.9 Gt per year, respectively, in the Base Case. *(See Figure A-1.)*

In this scenario, an array of policy measures (representing a mix of fiscal, tax, and other incentives) is implemented to spur the use of less CO_2-intensive fuels and to support the development of more efficient energy technologies. The efficiency of natural gas and nuclear energy supply is assumed to increase faster than in the Base Case, and energy end-use efficiency is assumed to grow more rapidly (1.0 percent per year compared to 0.8 percent in the Base Case). Penalties for the environmental impacts of coal and shale use are assumed to increase. Improvements in the efficiency of coal and shale oil supply occur at the same rates achieved in the Base Case. Lower estimated costs of solar energy in this scenario (i.e., $15.00 to $16.50 per EJ) reflect the assumption that stronger pro-solar policies will be adopted. At these lower prices, solar technologies are introduced more rapidly than in the Base Case.

As a consequence of the overall strategy assumed in this scenario, coal use grows more slowly than in the Base Case—by approximately 1.3 percent per year from 1975 to 2025 and 1.6 percent thereafter, compared to 2.4 percent per year to 2025 and 1.9 percent per year from 2025 to 2075 in the Base Case. Coal captures 34 percent of the market for primary energy in 2025 and 51 percent in 2075, compared to 47 percent in 2025 and 66 percent in the Base Case. *(See Figure 10.)* Nuclear energy use reaches approximately the same levels (about 20 EJ per year in 2025) as in the Base Case. However, because the total demand for primary energy is lower in the Modest Policies case, nuclear energy's market share is somewhat higher. Solar technology makes a slightly larger contribution to energy supply in the Modest Policies Scenario than it does in the Base Case. Solar systems contribute approximately 3 percent of primary energy in 2025 and 4.2 percent in 2050, compared to 2.5 percent and 4.0 percent, respectively, in the Base Case.

The biotic contribution to CO_2 is limited by a strong

global effort to reduce the rate of tropical deforestation and to reforest areas where trees once grew. The Tropical Forest Action Plan developed by the World Resources Institute, the UN Development Programme, the World Bank, and others is implemented during the late 1980s and early 1990s to stem forest conversions.[40] Advanced techniques for using fuelwood and other bio-energy sources more efficiently are introduced.[41] Investment strategies similar to that outlined in the tropical forest action plan are continued on a multilateral basis into the 21st century. As a result, the net biotic release of CO_2 declines from 1.0 Gt of carbon per year in 1975 to 0.8 Gt annually in 2025. The net annual biotic release falls to 0.65 Gt in 2075.

D. The Slow Build-up Scenario

In this scenario, strong global efforts to reduce greenhouse gas emissions eventually stabilize the atmosphere's composition. *(See Table 3.)* As a result, coal prices increase by 2.3 times between 1975 and 2075. Oil and gas prices increase by factors of 3.1 and 2.8 over the same period. *(See Table A-5.)* Per capita energy use *declines* in the industrialized countries by 0.37 percent per year and in the developing countries by 0.47 percent per year. *(See Table A-6.)* Total primary energy use grows slowly, from approximately 250 EJ per year in 1975 to approximately 260 EJ per year in 2075. *(See Figure 5.)* Primary energy use increases at an average rate of approximately 0.3 percent per year from 1975 to 2075, compared to 1.3 percent per year for the same period in the Base Case. *(See Figure 11.)*

CO_2 emissions from fossil fuel combustion decline from 4.5 Gt of carbon per year in 1975 to 2.0 Gt of carbon per year in 2075. *(See Figure A-1.)* This rate of decrease, approximately 0.8 percent per year, compares with a rate of emissions growth of about 1.3 percent per year in the Base Case.

This decrease in CO_2 emissions comes about because new policies stimulate higher efficiency in energy use at a rate of 1.5 percent per year. Lowering CO_2 emissions to this extent also requires governments and international organizations, for example, to stimulate rapid improvements in the efficiency of energy supply from the less CO_2-intensive energy systems. Thus, the efficiency of gas supply is assumed to increase by 1.5 percent per year (compared to 0.3 percent per year in the Base Case), fission energy supply efficiency grows by 0.4 percent per year (compared to 0.2 percent per year in the Base Case), and the efficiency of oil supply increases by 0.6 percent per year (compared to 0.3 percent per year in the Base Case). *(See Table 4.)*

Policy measures required to speed the introduction of solar energy (bringing prices to their minimum levels in 25 vs. 50 years) might include substantially increased research support, tax incentives, and other subsidies. Thanks to such initiatives, solar energy costs fall to only $12.00 per GJ (in constant 1975 dollars) in 2000, compared to a minimal cost of $16.50 GJ reached in 2025 in the Base Case and $15.00 per GJ reached in 2025 in the Modest Policies Case. Strong global efforts to limit the conversion of forests and other unmanaged or lightly managed ecosystems are also assumed.

By contrast, the efficiency of energy supply from coal and unconventional oil grows substantially more slowly than in the Base Case. The development of synfuels from coal is slowed by the lack of government support, reaching a minimum cost of $7.00 per GJ for synthetic oil and $5.50 per GJ for synthetic gas in 2035. This compares to minimal prices of $4.25 per GJ and $3.15 per GJ, respectively, achieved in the Base Case in 2005.

Total coal use in this scenario declines by approximately a factor of four between 1975 and 2075. Coal use in 2025 is limited to 19 EJ per year, compared to almost thirteen times that amount in the Base Case. Coal use remains relatively constant for the next fifty years in this scenario. As a result, in 2075, coal use is only 18 EJ per year, compared to 620 EJ per year in the Base Case.

The fuel mix in 2025 is significantly different from that in 1975. (Compare Figure 11 and Figure 8.) Natural gas provides 24 percent of total primary energy in 2025,

compared to 11 percent in the Base Case. In 2075, gas supplies 22 percent of primary energy, compared to about 5 percent in the Base Case. Solar technologies supply 6 percent of primary energy in 2025 and 7.7 percent in 2075, compared to 2.6 percent and 4.1 percent, respectively in the Base Case. Hydro contributes a larger share of primary energy in the Slow Build-up Scenario, but the total supply equals less than half of the hydro energy supplied after 2000 in the Base Case.

The biotic contribution in the Slow Build-up Scenario declines steadily over these one hundred years from 1 Gt per year in 1975 to 0.01 Gt in 2075. This substantial reduction in the net biotic release occurs because of massive reforestation programs, strong controls on deforestation, and the extensive and rapid introduction of more efficient end-use energy devices (such as improved wood stoves) in developing countries' non-commercial sectors. (The effects of such measures are not quantified here, however.)

Energy use and CO_2 emissions increase in three of the four WRI scenarios. In the Slow Build-up Scenario, energy stays nearly constant but CO_2 emissions decline by more than fifty percent.

V. Atmospheric Build-up of CO$_2$ and Other Greenhouse Gases

The differences in the emissions levels associated with each of the four scenarios described in Section IV translate into significant differences in future concentrations of greenhouse gases in the atmosphere. As the following comparisons of the scenarios show, these concentrations vary by factors of two to five among the four scenarios.

A. Carbon Dioxide

In the Base Case Scenario, the concentration of CO$_2$ reaches approximately 600 ppmv in about 2075, compared with approximately 820 ppmv in the High Emissions Scenario or 420 ppmv in the Slow Build-up Scenario. *(See Figure 12.)* The year in which atmospheric CO$_2$ reaches 550 ppmv varies from about 2040 for the High Emissions case to about 2070 in the Base Case and to well beyond 2075 in the Modest Policies and Slow Build-up Scenarios.

For perspective, these future atmospheric CO$_2$ concentrations were compared with the estimates made in four other recent studies. *(See Table 6.)* Nearly all of the projections reported in these earlier studies fall between WRI's High Emissions and Slow Build-up Scenarios.

The CO$_2$ build-up in the atmosphere over the next century varies by a factor of about three in these earlier studies. The atmospheric concentration projected for 2075 in the earlier studies ranges from 490 ppmv in the 5th percentile scenario of Nordhaus and Yohe to 1550 ppmv at the high end of the range for Edmonds and Reilly's Scenario A. By comparison, the WRI cases suggest a range of CO$_2$ concentrations of 420-820 ppmv.

B. Nitrous Oxide

In the WRI model, the rate of growth in N$_2$O emissions is assumed to equal the rate of growth in coal

use. *(See Figure 13.)* As Table A-7 shows, the demand for coal in 2025 varies by a factor of twenty, from approximately 390 EJ per year in the High Emissions Scenario to 20 EJ per year in the Slow Build-up case. By 2075, annual coal use in these two scenarios differs by a factor of 60.

Table A-10 illustrates the rate of growth in coal use in the WRI Scenarios. Figure 14 illustrates the resulting concentration of atmospheric N$_2$O. Estimated concentration in 2050 ranges from 445 ppbv in the High Emissions case to ˜310 ppbv in the Slow Build-up Scenario. For 1980 to 2050, the scenarios analyzed suggest growth rates for atmospheric N$_2$O concentration of from 0.05 percent per year to 0.56 percent per year. As Table 7 indicates, the estimates generated in this study are somewhat conservative but span most of the range indicated by scenarios developed in the six other recent analyses. The range of estimates for 2050 in these earlier studies runs from 360 ppmv in the Time-Dependent Scenarios of the WMO study (1986a) to 470 ppmv suggested by the combustion source model of Weiss (1981).

C. Chlorofluorocarbons

Table 8 illustrates the estimated global production of CFC-11 and CFC-12 in each scenario. Estimated production in 2075 varies by a factor of 12, from 750,000 metric tons in the Slow Build-up Scenario to 9,100,000 metric tons in the High Emissions case. Between 1980 and 2075, annual growth in production of these compounds ranges from about 2.6 percent per year in the High Emissions case to zero in the Slow Build-up Scenario.

Table 9 describes the assumptions made about the rate of growth in demand for CFCs in each region as income rises. The High Emissions scenario assumes that people in other regions increase their demand for products incorporating CFCs as rapidly as Americans

Figure 12. Atmospheric CO_2 Concentration in the WRI Scenarios (Parts per Million by Volume)

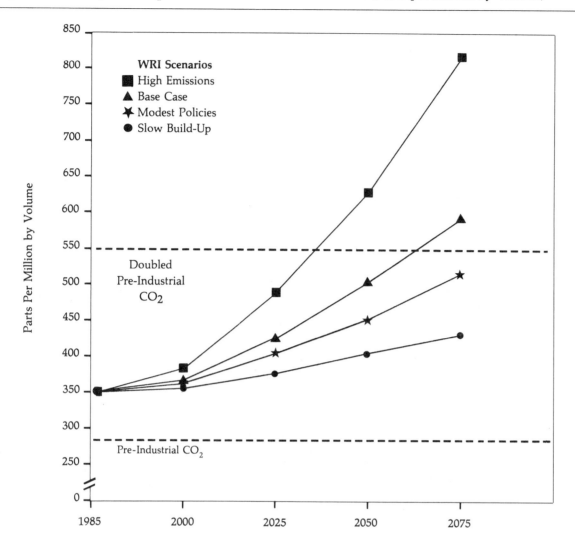

have in the last four decades. Thus, the future income elasticities of demand outside the United States resemble those that the United States experienced when U.S. markets were developing. In the Base Case Scenario, future income elasticities in the Eastern Bloc countries and in the developing countries are assumed to be significantly less than those encountered in the United States between 1940 and 1980. This assumption reflects the belief that these societies will invest significantly less in aerosols, flexible foams, and food packaging materials using CFCs than was the case in the United States.

The Modest Policies and the Slow Build-up Scenarios simulate the effects of international policies to limit global production and to control key end-uses of CFC-11 and CFC-12. In the Modest Policies case, an international agreement limits production to what global capacity was in 1985 (approximately 1,240,000 metric tons). In the Slow Build-up Scenario, production

is limited to the amount actually manufactured in 1985 (an estimated 750,000 metric tons per year). Furthermore, in the Modest Policies case, an international agreement mandates a 50-percent reduction in disposal losses in refrigeration applications beginning in 1990 (compared to the levels expected in the absence of policy intervention), a 30-percent decrease in manufacturing losses during foam-production beginning in 1995, and a shift of 50-percent of CFC production from foam-blowing to refrigeration applications in 1995. A similar hypothetical international agreement in the Slow Build-up Scenario stipulates that disposal losses in refrigeration applications are gradually reduced by 50 percent in 1990, by 75 percent in 2010, and by 90 percent in 2030; manufacturing losses in foam-blowing operations are reduced by 30 percent in 1995, by 50 percent in 2005, by 70 percent in 2015, and by 90 percent in 2025; and CFC use in foam applications is reduced by 50 percent in 1990, by 75 percent in 2010, and 90 percent

Figure 13. Coal Use in the WRI Scenarios (Exajoules per Year)

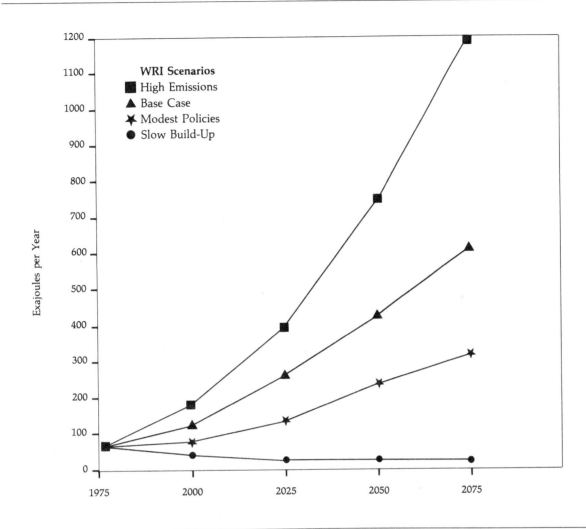

in 2030, with the unused production transferred to refrigeration applications.

What results do these assumptions yield? As Figures 15 and 16 show, annual emissions of CFC-11 vary among the scenarios by a factor of 10 in 2050, from 250,000 metric tons in the Slow Build-up Scenario to 2,600,000 metric tons in the High Emissions Scenario. Similarly, annual emissions of CFC-12 vary by a factor of about 8, from 300,000 metric tons to 2,300,000 metric tons in 2050. In Table 10, these estimates are compared to three other recent projections of future CFC emissions growth rates. As Table 11 shows, the estimated concentrations of CFC-11 in 2075 vary from 0.8 ppbv in the Slow Build-up Scenario to almost 6 ppbv in the High Emissions Scenario. The estimated concentration of CFC-12 in 2075 ranges from 1.8 ppbv in the Slow Build-up Scenario to about 7 ppbv in the High Emissions Scenario. As Table 12 indicates, the estimates based on the WRI scenarios suggest somewhat lower

rates of growth in atmospheric concentration than do most of the earlier studies.

D. Methane

Since the processes that control emissions and the removal of atmospheric methane are poorly understood, the model computes future concentrations based on arbitrary assumptions of growth rates. These vary from an upper limit of about 2 percent per year in the High Emissions Scenario to a lower limit of approximately 0.5 percent per year in the Slow Build-up Scenario, compared to the 1 percent annual rate of growth over the past 30 years. The projected range for observed methane concentrations in 2030 is between 2.1 and 4.6 parts per million by volume. *(See Table 13)*. The methane growth rates in these four scenarios produce concentrations in 2030 that span the range estimated in five earlier studies.

Figure 14. N$_2$O Concentration in the WRI Scenarios (Parts per Billion by Volume)

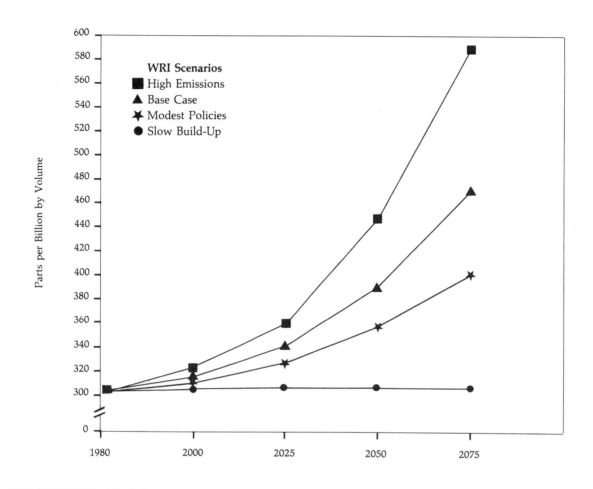

E. Tropospheric Ozone

The rate of growth projected for tropospheric ozone in this study is similar to the rates projected by Ramanathan *et al.* and by Bolin *et al.*, though the time horizon of these other studies was generally shorter than that considered here. *(See Table 14.)* In each WRI scenario, the concentration of tropospheric ozone increases linearly, reaching a maximum of about 15 percent above present levels in 2040. This assumption implies a rate of growth in tropospheric ozone concentration of approximately 0.25 percent per year.

Table 6   Projections of Atmospheric Carbon Dioxide Concentration

Scenario	1985	2000	2025	2050	2075
WRI					
High Emissions	345	381	482	627	817
Base Case	345	370	420	494	595
Modest Policies	345	368	408	457	518
Slow Buildup	345	362	383	403	419
NASA/WMO (a)	345	390	440	500	570
Nordhaus/Yohe (b)					
95th Percentile	345	420	540	800	1250
50th Percentile	345	390	460	550	625
5th Percentile	345	360	400	440	490
Edmonds-Reilly (c)					
Scenario A	345	360–400	480–530	760–840	1400–1550
Scenario B	345	380–400	440–480	540–600	670–760
Scenario C	345	370–400	420–460	470–520	510–580
Seidel and Keyes (d)					
Mid-Case	345	380	450	540	710

Notes:

(a) Future concentration estimates assume an annual increase of 0.5% in atmospheric concentration of CO_2. Source is NASA, ''Present State of Knowledge of the Upper Atmosphere,'' Reference Publication No. 1162, January, 1986.

(b) Future concentrations are approximate based on interpolation of Figure 2.4, p. 95, in Nordhaus and Yohe, ''Future Carbon Dioxide Emissions from Fossil Fuels,'' in NAS, ''Changing Climate,'' 1983.

(c) Data are drawn from Table 10.7, p. 271, in Trabalka, *et al.*, ''Human Alterations of the Global Carbon Cycle and the Projected Future,'' pp. 248–287, in J. Trabalka, ed., ''Atmospheric Carbon Dioxide and the Global Carbon Cycle'', Report No. DOE/ER-0239, December 1985.

(d) S. Seidel and D. Keyes, ''Can We Delay a Greenhouse Warming,'' U.S. EPA, September 1983.

Table 7. Estimated Future Concentrations of Nitrous Dioxide
(Parts Per Billion by Volume)

Scenario	1990	2010	2030	2050
WRI				
High Emissions	310	330	370	445
Base Case	305	320	350	390
Modest Policies	305	315	335	355
Slow Build-up	305	305	305	305
WEISS (1981)				
Fossil Fuel Source	310	330	380	470
WANG AND MOLNAR				
(1985)	305	320		
RAMANATHAN *et al.*				
(1985)				
Range			350–450	
Best Estimate			375	
MacDONALD (1985)			350–450	
DICKINSON & CICERONE				
(1986)			350–450	
NASA (1986)				
Time-Dependent	310	325	340	360

Sources:
R.E. Dickinson and R.J. Cicerone, ''Future Global Warming from Atmospheric Trace Gases,'' Nature, Vol. 319, January 9, 1986, pp. 109–115.
G.J. MacDonald, ''Climate Change and Acid Rain,'' Report No. MP86W00010, Mitre Corporation, McLean, VA., November 1985.
NASA, ''Present State of Knowledge of the Upper Atmosphere,'' Reference Publication No. 1162, January 1986.
V. Ramanathan, *et al.*, ''Trace Gas Trends and Their Potential Role in Climate Change,'' Journal of Geophysical Research, Vol. 90, No. D3, pp. 5547–5566, June 20, 1985.
W.O. Wang and G. Molnar, ''A Model Study of the Greenhouse Effects Due to Increasing Atmospheric CH_4, N_2O, CFC11, and CFC12,'' Journal of Geophysical Research, Vol. 90, No. D7, pp. 12971–12980, December 20, 1985.
R.F. Weiss, ''The Temporal and Spatial Distribution of Tropospheric Nitrous Oxide,'' Journal of Geophysical Research, Vol. 86, No. C8, pp. 7185-7195, August 20, 1981.

Table 8. Projected Total Production of CFC-11 and CFC-12
(Thousands of Metric Tons Per Year)

Year	Scenarios			
	High Emissions	Base Case	Modest Policies	Slow Build-Up
1990	1100	1100	1100	750
2000	1600	1200	1200	750
2010	2100	1300	1200	750
2020	2600	1400	1200	750
2030	3200	1500	1200	750
2040	4000	1600	1200	750
2050	5400	1800	1200	750
2075	9100	2100	1200	750
Annual Growth Rate 1985–2075	2.6%	0.95%	0.35%	0.00%

Table 9.   Assumed Growth Rates in Demand for CFC-11 and CFC-12

Scenario	MARKET DEVELOPMENT PERIOD		
	Early Growth Period (1990–2000)	Emerging Market (2000–2040)	Mature Market (2040–2075)
HIGH EMISSIONS (1)			
Western Market Economies	GPC Rate (4)	GPC Rate	GPC Rate
USSR and East Europe	Elasticity is at historic US rate	Elasticity is at historic US rate	GPC Rate
Developing Countries	Elasticity is at historic US rate	Elasticity is at historic US rate	75% of GPC Rate
BASE CASE (2)			
Western Market Economies	50% of GPC Rate	50% of GPC Rate	50% of GPC Rate
USSR and East Europe	75% of GPC Rate	75% of GPC Rate	75% of GPC Rate
Developing Countries	GPC Rate	Elasticity is half historic US rate	50% of GPC Rate
MODEST POLICIES (2)			
Western Market Economies	50% of GPC Rate	50% of GPC Rate	50% of GPC Rate
USSR and East Europe	75% of GPC Rate	75% of GPC Rate	75% of GPC Rate
Developing Countries	GPC Rate	Elasticity is half historic US rate	50% of GPC Rate
SLOW BUILD-UP (3)			
World Total	Production constant at 1985 levels	Production constant at 1985 levels	Production constant at 1985 levels

Notes:

1. Production for aerosol uses of CFC-11 and CFC-12 in the U.S. are assumed to be zero after 1980. Production for non-aerosol uses in the U.S. assumed to grow at 7.2% per year until 1990 after which the growth rate is expected to fall by 2010 to a level equivalent to the rate of growth in GNP per capita (the GPC rate). In the other OECD countries, growth rate for aerosol production is assumed to be zero after 1980. The growth rate of non-aerosol uses is assumed to be 3.6% per year until 1990 in other OECD countries and to fall to the GPC rate by 2010. Production for aerosol uses in the East Bloc are assumed to grow at 5.2% per year to 1990 and then to increase with an income elasticity equal to historic US rates. In 2020, growth rates decline to the GPC rate. Non-aerosol production in the East Bloc grows at 5.2% per year until 1990 then increases with an income elasticity equal to historic U.S. rates. In 2020 growth slows to the GPC rate. In the developing countries, production grows at the GPC rate until 1990 when elasticities rise to historic U.S. levels. In 2060, growth rates decline in the developing countries to 75% of the GPC rate.
2. Production for aerosol uses of CFC-11 and CFC-12 in the U.S. are assumed to be zero after 1980. In the other OECD countries the growth rate in production for aerosol uses is assumed to be zero after 1980. The growth rate of non-aerosol production of CFC-11 is assumed to be 7.2% per year until 1990 in the OECD countries and then to fall to half the GPC rate. The growth rate of non-aerosol production of CFC-12 is assumed to be 3.6% per year until 1990 in the OECD countries and then to fall to half the GPC rate. Production of CFCs for aerosol uses in the USSR and the East Bloc is assumed to grow at 5.2% per year until 1990 and at 75% of the GPC rate after that. Production for non-aerosol uses of CFC-11 is expected to grow at 7.2% per year until 1990 and at 75% of the GPC rate from 1990 to 2075. Production of CFC-12 for non-aerosol applications is assumed to grow at 3.6% per year to 1990 and then to decline to 75% of the GPC rate in the USSR and the East Bloc. In the developing countries, production of CFC-11 and CFC-12 is assumed to occur at the GPC rate from 1980 to 2000. From 2000 to 2040, production is expected to grow at a rate equal to half the historic U.S. elasticities times the GPC rate. After 2040, the rate of growth in production falls to half the GPC rate in developing countries.
3. Global production of CFC-11 and CFC-12 remains constant at the estimated 1985 level from 1985 to 2075 in these scenarios.
4. GPC = GNP per capita.

Figure 15. CFC-11 Concentration in the WRI Scenarios (Parts per Billion by Volume)

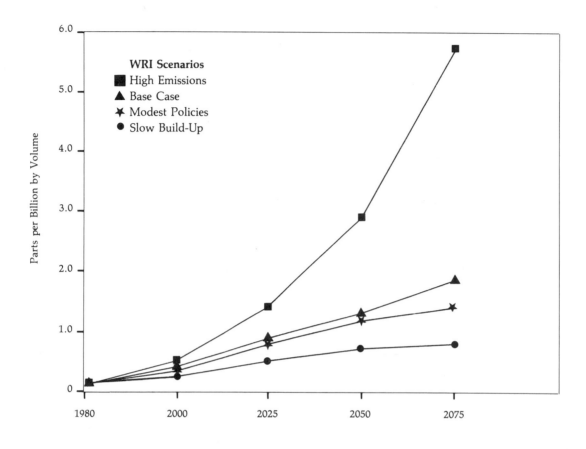

Sources for Table 10:

F. Camm *et al.*, ''Joint Emissions Scenarios for Potential Ozone Depleting Substances,'' Paper prepared for the U.S. Environmental Protection Agency Workshop on Demand and Control Technologies to Protect the Ozone Layer, Report No. WD-2942-1-EPA, February 1986.

T.H. Quinn *et al.*, ''Projected Use, Emissions, and Banks of Potential Ozone-Depleting Substances,'' RAND Report No. N-2282-EPA, January 1986.

U.S. Environmental Protection Agency, ''Analysis of Growth in Markets for Potential Ozone-Depleting Chemicals: 1985–2000,'' Draft Summary Paper prepared for U.S. EPA Workshop on Demand and Control Technologies to Protect the Ozone Layer, March 6–7, 1986, Washington, D.C.

Table 10. Projected Growth Rates in CFC Emissions

Scenario	CFC-11	CFC-12
WRI: This Study 1980–2075		
High Emissions	3.0%	2.3%
Base Case	1.2%	1.0%
Modest Policies	0.8%	0.2%
Slow Build-Up	−0.1%	−0.4%
Quinn *et al.* (1986): 1980–2075		
Late Market Maturation	2.7%	2.3%
Late Market Maturation/EEC Cap	2.6%	2.2%
East Bloc/LDC Growth	2.3%	2.1%
Early Market Maturation	2.1%	1.8%
Slow Growth	1.5%	1.3%
Technological Displacement	0.8%	1.3%
U.S. EPA (1986):		
1985–2000	0–6.0%	(−0.4)–4.9%
Camm *et al.* (1986):		
1985–2075		
5th Percentile	0.3%	0.8%
50th Percentile	1.5%	1.4%
95th Percentile	2.8%	2.6%

Figure 16. CFC-12 Concentration in the WRI Scenarios (Parts per Billion by Volume)

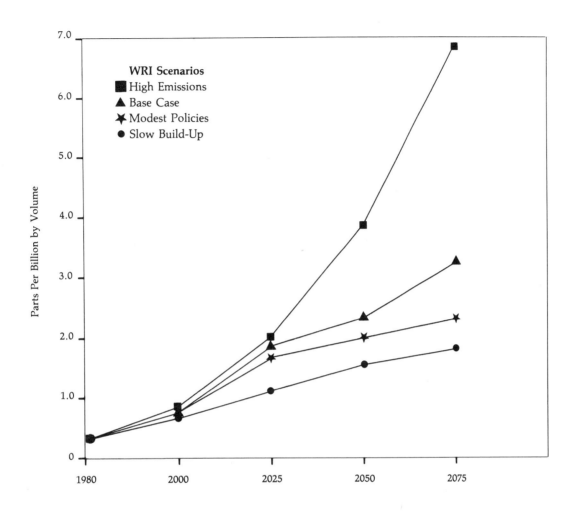

Table 11. Growth Rates in CFC Concentration in the WRI Scenarios
(All growth rates given in percent per year)

Scenario	CFC-11	CFC-12
High Emissions	3.6% per year	3.2% per year
Base Case	2.4	2.4
Modest Policies	2.3	2.2
Slow Build-Up	1.4	1.7

Table 12. Comparison of WRI Estimates of Future CFC Concentrations with Other Published Studies
(All Concentrations in ppbv)

Source of Estimate	2030 CFC-11	2030 CFC-12	2050 CFC-11	2050 CFC-12
WRI: This Study	0.6-1.7	1.2-2.3	0.7-2.9	1.5-3.8
Dickinson and Cicerone (1986)			0.7-3.0	2.0-4.8
MacDonald (1985)	0.5-1.0	0.6-1.2		
Ramanathan *et al.* (1985)	0.5-2.0	0.9-3.5		

Table 13.	Estimated Future Concentrations of Methane (Parts Per Million by Volume)			
Scenario	**1990**	**2010**	**2030**	**2050**
WRI				
High Emissions	1.78	2.77	4.55	7.45
Base Case	1.73	2.12	2.58	3.15
Modest Policies	1.73	2.04	2.37	2.75
Slow Build-Up	1.69	1.87	2.07	2.22
Wang and Molnar (1985)	1.90	2.40		
Ramanathan *et al.* (1985)				
Range			1.85-3.30	
Best Estimate	1.80	2.10	2.34	
MacDonald (1985)			2.50-3.30	
Dickinson & Cicerone (1986)				2.1-4.0
NASA (1986)				
Time Dependent	1.80	2.20	2.70	3.30

Sources for Table 12 and Table 13:
R.E. Dickinson and R.J. Cicerone, ''Future Global Warming from Atmospheric Trace Gases,'' Nature, Vol. 319, January 9, 1986, pp. 109–115.

G.J. MacDonald, ''Climate Change and Acid Rain,'' Report No. MP86W00010, Mitre Corporation, McLean, VA., November 1985.

NASA, ''Present State of Knowledge of the Upper Atmosphere,'' Reference Publication No. 1162, January 1986.

V. Ramanathan, *et al.*, ''Trace Gas Trends and Their Potential Role in Climate Change,'' Journal of Geophysical Research, Vol. 90, No. D3, pp. 5547–5566, June 20, 1985.

W.C. Wang and G. Molnar, ''A Model Study of the Greenhouse Effects Due to Increasing Atmospheric CH_4, N_2O, CFC11, and CFC12,'' Journal of Geophysical Research, Vol. 90, No. D7, pp. 12,971–12,980, December 20, 1985.

Table 14.	Estimated Growth Rate in the Concentration of Tropospheric Ozone (Percent per year)
Study	**Estimated Rate of Growth**
WRI: This Study	0.23% until 2040
Ramanathan *et al.* (1985)	0.25% until 2030
MacDonald (1985)	0.71-0.81% until 2030
Bolle, Seiler & Bolin (1985)	0.25% below 9 km
Dickinson & Cicerone (1986)	0.21-0.62% below 12 km

Sources:
H.J. Bolle, W. Seiler, and B. Bolin, ''Other Greenhouse Gases and Aerosols'' in B. Bolin, ed. ''The Greenhouse Effect, Climate Change, and Ecosystems,'' Wiley and Sons, 1986.

R.E. Dickinson and R.J. Cicerone, ''Future Global Warming from Atmospheric Trace Gases,'' Nature, Vol. 315, January 9, 1986, pp. 109–115.

G.J. MacDonald, ''Climate Change and Acid Rain,'' Report No. MP86W00010, Mitre Corporation, McLean, VA., November 1985.

V. Ramanathan, *et al.*, ''Trace Gas Trends and Their Potential Role in Climate Change,'' Journal of Geophysical Research, Vol. 90., No. D3, pp. 5547–5566, June 20, 1985.

VI. Modelling Results and Policy Implications

The warming effects of these four illustrative scenarios of greenhouse gas build-up, provides a framework for evaluating the potential of policy to limit future global warming. The "atmospheric commitment" to the so-called equilibrium warming is emphasized because once greenhouse gases are emitted into the atmosphere, the changes are essentially irreversible; no known policy can rapidly remove them.

Tables 15 to 18 and Figure 17 depict the progress of global warming in the four WRI scenarios. Tables 15 to 18 show the direct warming attributable to each of the six gases and the equilibrium warming that their combined effects will eventually cause. The greenhouse

---

Once greenhouse gases are emitted into the atmosphere, the changes are essentially irreversible; no feasible policy can rapidly remove them.

---

gas concentrations on which these estimates are based are summarized in Appendix Tables A-11—A-14.

Depending on which policies are adopted, the year when we are irreversibly committed to a warming of 1.5° to 4.5°C above the pre-industrial temperature varies by approximately six decades. In the Base Case Scenario this point is reached before 2030. In the Slow Build-up Scenario, an equivalent warming is postponed until after 2075, compared to about 2015 in the High Emissions Scenario. N_2O and tropospheric ozone contribute relatively less to this difference of sixty years than do CO_2, CFCs, and methane.

By 2075, the differences in atmospheric concentrations and warming commitments among the four scenarios are even more pronounced. In the High Emissions Scenario, by 2075 the build-up of greenhouse gases commits Earth to an equilibrium warming of

5.3°-16.0°C above the pre-industrial temperature. In the Slow Build-up Scenario, by contrast, the total commitment to global warming is limited to approximately 1.4°-4.2°C. For the Base Case, the total commitment is 2.9°-8.6°C in 2075, and, for the Modest Policies Case, 2.3°-7.0°C. Clearly, the magnitude and

---

Depending on which policies are adopted, the year when we are irreversibly committed to a warming of 1.5° to 4.5°C above pre-industrial temperature varies by approximately six decades.

---

timing of planetary warming can be substantially affected by policy choices made now and implemented over the next several decades.

The Base Case scenario reflects conventional wisdom in its assumptions about technological change, economic growth, and the evolution of the global energy system from 1980 to 2075. No new policies are implemented to slow the rate of greenhouse gas emissions. No major effort is made to retard tropical deforestation or to make energy use more efficient. Environmental concerns play no part in energy policy. This scenario is predicated upon conservative (lower-than-historical) rates of increase in CFC production and emissions, and yet the model projects a commitment to an average global warming of 0.9°-2.6°C above the pre-industrial level by 2000. By 2030, without changes in energy policy, the combined effects of the major greenhouse gases would lead to a commitment to global warming of 1.6°-4.7°C, slightly more than would result from doubling the pre-industrial concentration of CO_2 alone.

This is by no means the worst possible outcome. If the use of coal and synfuels is encouraged, if the rate of

Table 15. Base Case Scenario Commitment to Equilibrium Warming
(Degrees Centigrade)

| Year | Radiative Forcing | | | | | Warming Commitment Relative to 1980 Atmosphere | Warming Commitment Relative to the Pre-Industrial Atmosphere |
	CO_2 Forcing	CO_2+N_2O Forcing	CO_2+N_2O+ CH_4 Forcing	CO_2+N_2O+ $CH_4+Ozone$ Forcing	CO_2+N_2O+ CH_4+CFCs $+Ozone$ Forcing		
1980	0.0	0.0	0.0	0.0	0.0	0.0-0.0	0.5-1.5
1990	0.1	0.1	0.1	0.1	0.2	0.2-0.5	0.7-2.0
2000	0.2	0.2	0.3	0.4	0.5	0.4-1.1	0.9-2.6
2010	0.4	0.4	0.5	0.6	0.8	0.6-1.7	1.1-3.2
2020	0.5	0.6	0.7	0.8	1.1	0.8-2.4	1.3-3.9
2030	0.7	0.8	1.0	1.1	1.4	1.1-3.2	1.6-4.7
2040	0.9	1.0	1.3	1.4	1.8	1.3-4.0	1.8-5.5
2050	1.1	1.2	1.5	1.6	2.1	1.6-4.8	2.1-6.3
2060	1.3	1.5	1.9	2.0	2.5	1.9-5.7	2.4-7.2
2075	1.7	1.9	2.4	2.5	3.2	2.4-7.1	2.9-8.6

Notes:

1. The estimated values for change in surface temperature are extrapolated from the values given in Ramanathan *et al.* (1985). The CO_2 temperature effect is scaled logarithmetically. The temperature effects of CH_4 and N_2O are scaled to the difference in the square roots of the concentration in the perturbed and reference (1980) atmospheres. The temperature effects of CFC-11, CFC-12, and tropospheric ozone are scaled linearly. The total change in surface temperature is calculated by scaling the sum of the direct warming effects by a range of feedback factors equal to 0.75-2.25. This has the effect of scaling the climate sensitivity from 2°C to a range of 1.5°-4.5°C given in NAS (1983) for doubled CO_2. Use of these feedback factors may overstate the equilibrium warming effect somewhat in the High Emissions scenario.

2. Estimated changes in tropospheric ozone assume a 15% increase by 2040 and no change thereafter. This assumption may understate the effects of changes in tropospheric ozone between 2040 and 2075.

3. Warming commitment refers not to the temperature change experienced in a given year, but rather to the warming to which the Earth would be committed due to the cumulative emissions to that date, with the change in temperature measured relative to the 1980 atmosphere.

Table 16. High Emissions Scenario Commitment to Equilibrium Warming
(Degrees Centigrade)

| Year | Radiative Forcing | | | | | Warming Commitment Relative to 1980 Atmosphere | Warming Commitment Relative to the Pre-Industrial Atmosphere |
	CO_2 Forcing	CO_2+N_2O Forcing	CO_2+N_2O+ CH_4 Forcing	CO_2+N_2O+ $CH_4+Ozone$ Forcing	CO_2+N_2O+ CH_4+CFCs $+Ozone$ Forcing		
1980	0.0	0.0	0.0	0.0	0.0	0.0-0.0	0.5-1.5
1990	0.1	0.1	0.2	0.2	0.2	0.2-0.5	0.7-2.0
2000	0.3	0.3	0.5	0.5	0.6	0.5-1.4	1.0-2.9
2010	0.6	0.6	0.9	0.9	1.1	0.9-2.6	1.4-4.1
2020	0.9	0.9	1.3	1.4	1.7	1.3-3.9	1.8-5.4
2030	1.2	1.3	1.8	1.9	2.4	1.8-5.5	2.3-7.0
2040	1.5	1.7	2.3	2.4	3.2	2.4-7.1	2.9-8.6
2050	1.8	2.0	2.9	3.0	4.0	3.0-8.9	3.5-10.4
2060	2.1	2.4	3.5	3.6	4.9	3.7-11.0	4.2-12.5
2075	2.6	3.0	4.5	4.6	6.4	4.8-14.5	5.3-16.0

Notes
See Notes to Table 15.

Table 17. Modest Policies Scenario Commitment to Equilibrium Warming
(Degrees Centigrade)

| Year | Radiative Forcing | | | | | Warming Commitment Relative to 1980 Atmosphere | Warming Commitment Relative to the Pre-Industrial Atmosphere |
	CO_2 Forcing	CO_2+N_2O Forcing	CO_2+N_2O+ CH_4 Forcing	CO_2+N_2O+ $CH_4+Ozone$ Forcing	CO_2+N_2O+ CH_4+CFCs $+Ozone$ Forcing		
1980	0.0	0.0	0.0	0.0	0.0	0.0-0.0	0.5-1.5
1990	0.1	0.1	0.1	0.1	0.2	0.1-0.4	0.6-1.9
2000	0.2	0.2	0.3	0.3	0.4	0.3-1.0	0.8-2.5
2010	0.3	0.4	0.5	0.5	0.7	0.5-1.5	1.0-3.0
2020	0.4	0.5	0.6	0.7	0.9	0.7-2.1	1.2-3.6
2030	0.6	0.6	0.8	0.9	1.2	0.9-2.7	1.4-4.2
2040	0.7	0.8	1.0	1.1	1.5	1.1-3.3	1.6-4.8
2050	0.9	0.9	1.2	1.3	1.7	1.3-3.9	1.8-5.4
2060	1.0	1.1	1.4	1.5	2.0	1.5-4.5	2.0-6.0
2075	1.2	1.4	1.7	1.8	2.4	1.8-5.5	2.3-7.0

Notes
See Notes to Table 15.

Table 18. Slow Build-Up Scenario Commitment to Equilibrium Warming
(Degrees Centigrade)

| Year | Radiative Forcing | | | | | Warming Commitment Relative to 1980 Atmosphere | Warming Commitment Relative to the Pre-Industrial Atmosphere |
	CO_2 Forcing	CO_2+N_2O Forcing	CO_2+N_2O+ CH_4 Forcing	CO_2+N_2O+ $CH_4+Ozone$ Forcing	CO_2+N_2O+ CH_4+CFCs $+Ozone$ Forcing		
1980	0.0	0.0	0.0	0.0	0.0	0.0-0.0	0.5-1.5
1990	0.1	0.1	0.1	0.1	0.2	0.1-0.4	0.6-1.9
2000	0.2	0.2	0.2	0.2	0.3	0.3-0.8	0.8-2.3
2010	0.2	0.2	0.3	0.4	0.5	0.4-1.1	0.9-2.6
2020	0.3	0.3	0.4	0.5	0.6	0.5-1.4	1.0-2.9
2030	0.4	0.4	0.5	0.6	0.8	0.6-1.7	1.1-3.2
2040	0.4	0.4	0.6	0.7	0.9	0.7-2.0	1.2-3.5
2050	0.5	0.5	0.6	0.7	1.0	0.8-2.3	1.3-3.8
2060	0.5	0.5	0.7	0.8	1.1	0.8-2.5	1.3-4.0
2075	0.6	0.6	0.8	0.9	1.2	0.9-2.7	1.4-4.2

Notes
See Notes to Table 15.

Figure 17. Commitment to Future Warming in the WRI Scenarios

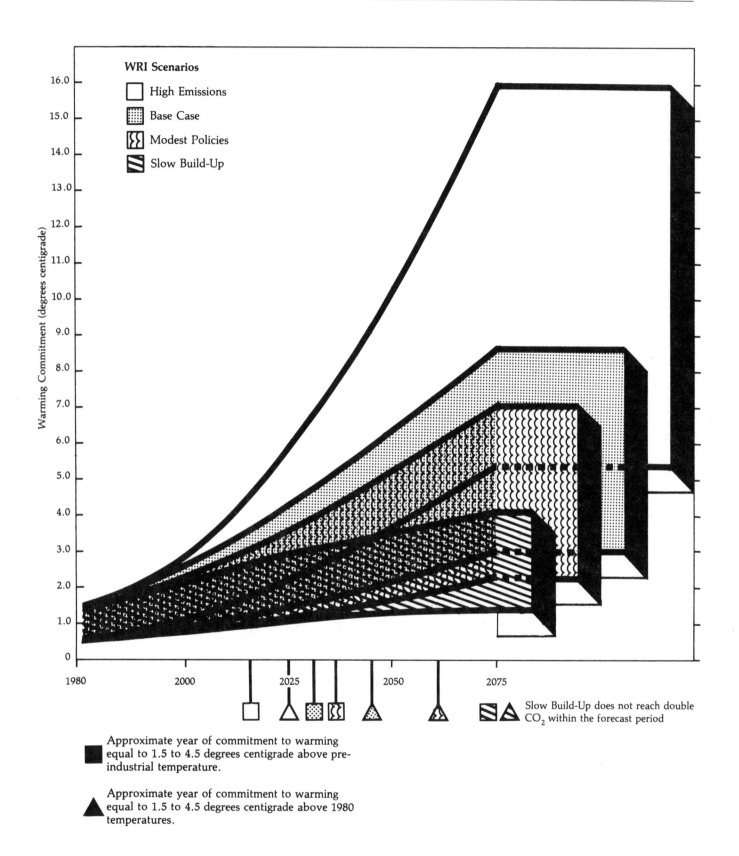

tropical deforestation is allowed to increase, if energy prices continue to ignore the environmental costs of energy supply and use, and if world use of CFC-11 and CFC-12 grows with incomes at close to the historical rate experienced in the United States, then even more dramatic changes in climate are probably in store. Under such a high-emissions scenario, the commitment to global warming could reach 1-3 °C over pre-industrial levels by 2000. If allowed to continue, such a scenario would commit the planet to an increase of 2.3°-7.0°C over the pre-industrial climate by 2030.

Although the planet is not locked in to either of these hot house futures, we no longer appear to have the opportunity to avoid a substantial greenhouse warming altogether. Under the Slow Build-up Scenario modelled here, for instance, energy policies hypothetically implemented in 1980 lead to rapidly improved energy

---

Fortunately, the planet is not ''locked in'' to ''hot house'' futures.

---

efficiency, greater use of solar energy systems, less reliance on solid fuels over the long term, and slowed tropical deforestation. In addition, international agreements cap CFC production at the 1985 level, and emissions rates are reduced through more careful management of CFC-using technologies. Even with these policies, if the model is correct, the global commitment to greenhouse warming would be 1.1°-3.2°C above pre-industrial levels in 2030, and 1.4°-4.2°C above pre-industrial levels in 2075.

These scenarios make it clear that any delay in making policy choices carries significant consequences. Indeed, even a heroic global effort may not be able to forestall a warming greater than any experienced during recorded human history. If societies wait several decades hoping scientists will penetrate all the chemical and physical workings of the atmosphere, the global warming that we and our children will live with is likely to be much larger than need be. The Model of Warming Commitment results indicate that *if 30 years of delay is allowed for removing scientific uncertainties, identifying options, establishing international consensus, and implementing appropriate policies, Earth will be committed to a warming 0.25°-0.8°C higher than that which would occur if the policies envisioned in the Slow Build-up Scenario were implemented today.* This increase—though a mere fraction of one degree—equals nearly 50 percent of the total emissions deposited in the atmospheric ''bank'' through human activities and biotic processes in the 100 years since the Industrial Revolution.

In any case, the inevitable legacy of today's industrial activities will be future climate changes, and the extent of those changes will be largely determined by human choices. In the WRI scenarios, the equilibrium warming

to which the atmosphere will be committed in 2075 varies by more than a factor of four. The estimated emissions in the High Emissions and Base Case Scenarios will commit the atmosphere to an eventual warming beyond that seen in the last million years.

In any case, the inevitable legacy of today's industrial activities will be future climate changes, and the extent of those changes will be largely determined by human choices.

Time is passing quickly. Atmospheric concentrations of greenhouse gases increase steadily. Significant climate change already may be unavoidable given current and historical rates of emissions. If governments conclude that a substantial further warming poses unacceptable social risks, then strong actions need to be taken, and taken soon. The longer the delay before preventive policies are identified, agreed upon, and implemented, the more extreme the policies imposed to stay within prudent bounds will have to be.

The results also suggest however that strategies do exist to control greenhouse gas build-up. Energy policies similar to the ones assumed in the Slow Build-up Scenario are needed to improve the efficiency of energy use, limit long-term commitments to solid fuels,

---

The estimated emissions in the High Emissions and Base Case Scenarios will commit the atmosphere to an eventual warming beyond that seen in the last million years.

---

and reduce the biotic contribution to CO_2 by slowing global deforestation. Such steps could reduce the future contribution to global warming from CO_2 by almost two thirds compared to the Base Case and by over 75 percent compared to the High Emissions Scenario. A ceiling on the production of CFC-11 and CFC-12 and controls on the uses of these chemicals could reduce the contribution to global warming from these compounds by over 60 percent compared to the Base Case and by over 80 percent compared to the High Emissions Scenario. Significantly reducing the cumulative contribution to warming from methane build-up would probably require controls on methane leakage from fossil fuel technologies and from biomass burning,

Unknowns, Uncertainties, and Critiques of the Models

Projecting future global warming due to emissions of greenhouse gases requires estimating the annual amount of energy used, the mix of fuels consumed, and the extent of activities that affect the sources and sinks of all other relevant gases. In turn, these estimates must be based on projections of future population and economic growth and assessments of the relationships among economic growth, energy use, agricultural practices, and demands for products and services associated with emissions of other greenhouse gases. Moreover, to translate the resulting emissions estimates into anticipated future concentration levels, the chemical and biological processes that remove some of the annual emissions from the atmosphere and deposit them in other reservoirs in the biogeosphere must be simulated. Converting the resultant concentration levels to estimates of future global warming requires a model that represents the physical processes determining the atmosphere's radiation budget.

A. Uncertainties in the Data and Assumptions

Significant uncertainties exist in the data used to project levels of economic activity and emissions of greenhouse gases. In particular, demographic uncertainties greatly influence projections of future levels of economic activity and energy use. Today, most scientists agree on the rates used here of population increase over the next century. These projections—roughly in line with the World Bank/ U.N. estimates—suggest that the global population will reach approximately six billion people in 2000 and ten billion in 2100. In their review of recent energy and CO_2 scenarios, however, Ausubel and Nordhaus[42] found estimates for rates of global population growth for the period 1975 to 2025 ranging from 1.2 percent to 1.7 percent per year, compared to the estimates of approximately 1.6 percent between 1975 and 2000, and 0.5 percent between 2000 and 2100, used in the mid-range estimates of the United Nations. Whatever the conventional wisdom of the day, however, future rates of population growth twenty-five or seventy-five years from now cannot be predicted with certainty.

Besides the absolute level of population in a given year, other demographic factors affect rates of economic growth and energy use. Urbanization is an important demographic factor that is not effectively measured in the estimated rates of future population growth or in assumptions about constant income elasticities of demand. In developing countries, urbanization can triple or quadruple per capita demand for primary energy as charcoal is substituted for fuelwood.[43] On the other hand, when kerosene, liquid petroleum gas, electricity, or other high-quality fuels are substituted for traditional fuels, urban households that can afford them can get by with using as little as 30 percent of the energy their rural counterparts require for cooking—the largest single household energy demand in most developing countries.[44]

Unfortunately, few global population estimates reflect the effects of regional changes in social and cultural values or national population policies. For example, its current "one family, one child" campaign has substantially lowered the birthrate in the People's Republic of China, where one quarter of humanity now lives. Such national policies affect a population's age structure and, as a consequence, future projections of labor force participation rates and the rate of growth in labor productivity. Any of these changes can influence energy use and economic growth.

Major economic variables also affect projections of the production and emissions of greenhouse gases. Recent analyses of future energy use and economic growth incorporate differing estimates of future rates of increase in regional GNP and in GNP per capita. Even when looking forward for only one decade and at only one or two factor prices, economic experts cannot reach consensus. In one recent survey of estimates for future global GNP and oil prices in the next 25 years, the 70 governmental and international agencies, corporations, research institutes, and universities gave estimates of GNP in the year 2010 that varied by a factor of four.[45] In comparison, a global climate modeller must look forward fifty to one hundred years—an impossibly long time frame for economic forecasts. When compounded over half a century, even a small difference in the estimated rate of growth significantly affects the global GNP projected for the final year of the forecast period.

An array of economic, institutional, and physical factors may affect the rate of market penetration of new energy technologies. Laurmann, building on work by Marchetti, suggests that new energy supply technologies require 50 years to capture half of the market.[46] Conditions for market penetration, however, differ with cultural and economic systems, local institutional and regulatory settings, and technological characteristics. Thus, using a single-value estimate to represent the introduction of solar or synfuels technologies in each scenario for all societies and all time periods is too simplistic.

Significant uncertainty also could undermine assumptions made about other important economic parameters. *(See Table 4 and accompanying notes.)* The scenarios analyzed in this study embody alternative assumptions about the income and price elasticities of demand for energy as well as about the income elasticity of demand for CFCs. Recent analyses by Williams *et al.* suggest that structural changes in the western market economies and saturation of certain key end-uses could bring long-run income elasticities of demand as low as 0.35 for OECD countries, compared to the values of 0.8 to 1.0 considered in this study.[47]

Various technological factors also affect the rate of growth of future emissions. How fast the efficiency of energy supply and use will improve, how emerging environmental constraints (other than global warming) might affect the rate at which energy technologies can be mobilized, and how fast technologies for producing and using CFCs might improve are all uncertain.

In addition, as yet unidentified feedback effects between climate change and future patterns of energy use or greenhouse gas emissions could come into play. Jaeger for example, has used two general circulation models to investigate the impacts of doubled CO_2 on the conditions that would affect energy demand in five West European cities.[48] Her results suggest that global warming could decrease the heating season by one to three months and increase the cooling season significantly—with commensurate impacts on demands for fuel and electricity. None of these impacts are considered in the Model of Warming Commitment.

Changes in the composition of energy demand and in the demand for CFCs can result from changes in production processes, the mix of economic activities, or values and lifestyles. For example, with the advent of computers and video communications links, long-distance transport needs may eventually be reduced. More generally, changing consumption habits, occupational preferences, and settlement patterns can increase or reduce per capita demand for energy and other goods.[49]

Edmonds and Reilly conducted a Monte Carlo analysis to determine which of the parameters considered in their model most affect the model's projections of energy demand and CO_2 emissions.[50] They found that the four most important input parameters vis-a-vis carbon emissions are 1) labor productivity in developing countries, 2) labor productivity in industrialized countries, 3) the rate of improvement in energy efficiency, and 4) income elasticity in developing countries. As for effects on

GNP, only the first two parameters, say the authors, explain significant variability in the projections of various scenarios. To explain uncertainty in world oil prices, Edmonds and Reilly find that a fifth parameter—the price-induced interfuel substitution in end-uses—is also important. Other significant but less important variables include the costs of producing energy from biomass, the environmental costs of coal extraction and use, income elasticity of demand for energy in the industrialized countries, aggregate price elasticity of demand for energy, and the rate of improvement in the efficiency of coal supply.

Considerable uncertainty remains in estimates of the biotic release of CO_2. Recent estimates suggest that the current net biotic contribution to atmospheric CO_2 is between -0.5 and 2.5 Gt of carbon per year.[51] As part of a recent research effort sponsored by WMO/ICSU/UNEP, the current rate of release was estimated to be 1 to 2 Gt of carbon per year.[52] In these scenarios, the assumption is that the net biotic release could reach a maximum of 7.5 Gt of carbon per year or a minimum of 0.01 Gt of carbon per year. In the Base Case Scenario, the assumption is that for the next 80 years the net biotic contribution will be 1 Gt of carbon per year. Little is known about the precise determinants of these rates or about the effects of feedbacks between global warming and future rates of CO_2 release from the biota. None of these feedback effects have been incorporated into the Model of Warming Commitment.

Substantial uncertainties also reside in the data and assumptions used to project future emissions of nitrous oxide and methane. In this study, the net contribution from biotic sources to future growth in release rates for N_2O is assumed to be the same in all four scenarios. Further research on the effects of deforestation and on soil nitrification and denitrification may prove this assumption false. Additional research and development of coal combustion technology may break the link between coal use and future N_2O emissions.

In the case of net increases in atmospheric methane, even less is known. Methane is released in a highly decentralized fashion by both biotic and anthropogenic sources. The emissions from each of these sources are not well documented. The situation is made more complex by uncertainty about the release rate for carbon monoxide—not a greenhouse gas, but a compound whose emissions affect the rate of increase in methane concentrations.[53]

Further uncertainties are present in the data and assumptions used in the models to convert future emissions to projected future concentrations of these gases. For CO_2, considerable uncertainty remains about the proportion of emitted CO_2 that remains in the atmosphere and about the atmospheric concentration before the Industrial Revolution. For N_2O, there is uncertainty about the steady-state pre-industrial concentration. In the case of tropospheric ozone, the processes and quantities determining the rate of build-up are poorly understood and incompletely defined. As these uncertainties are resolved by future research, the structure of the atmospheric retention models used in this study may have to be changed.

The uncertainties present in these data are compounded by the effects of uncertainties in estimates of the atmosphere's temperature sensitivity to the build-up of greenhouse gases. The assumed sensitivities for each gas may not completely reflect the interaction between these gases and the overlap of their absorption spectra—overlaps that could cause the model to overstate a gas's warming effects.

B. Critique of the Models

The models used in this analysis also have structural limitations. In the case of the energy-economic sub-model, the most outstanding weakness is that demand is not disaggregated. Only three end-use sectors in each region are used for the industrialized countries. For the developing countries, it is impossible in this model to simulate disaggregated improvements in energy efficiency since all end-uses are treated as a single aggregated activity.

Another limitation of the energy-economic model used here is its handling of the introduction of renewable technologies and synfuels. The model assumes that market penetration can be accurately simulated by a simple logistic function determined by the time needed to reach a breakthrough price and a fixed minimum cost of supply. The model has only minimal ability to represent the effects of policies to accelerate market penetration.

This energy-economic model has other structural weaknesses as well. Capital formation and depreciation are not incorporated into the model, so any simulation of policies affecting investment and facility construction is impossible. Changes in the mix of energy-supply technologies may be determined more by the effects of taxes and subsidies on investment decisions than by the effects of final energy prices on consumer decisions. Furthermore, because supply and demand are not determined primarily by price in centrally planned economies or where governments intervene actively in energy markets, the appropriateness of

using a partial-equilibrium model for analyzing the behavior of such economies is open to question.[54]

Other sub-models used in this study also have structural limitations. The simple model used to simulate future production of N_2O considers only the fossil fuel contribution to net releases of N_2O. For simplicity's sake, it also assumes that future emission growth rates will approximately equal the rate of growth in coal use. This approach ignores both the possible biotic contribution to N_2O emissions and the release of N_2O from combustion of fuels other than coal. As a result, annual emissions of N_2O may be understated.

The approach used here to estimate future CFC production is based on income elasticity of demand. Price effects on future CFC demand are not recognized in the model. The models used to translate future CFC production into estimated future emissions are constrained by their lack of regional disaggregation. Although the Model of Warming Commitment estimates future production of CFC-11 and CFC-12 on a regionally disaggregated basis, the analysis of particular end-uses and resulting prompt and banked emissions is aggregated globally, so it is difficult to accurately simulate the impacts of emissions-control policies implemented at a national or regional level.

Recent investigations with two-dimensional models have shown significant latitudinal and seasonal variation in the future distribution of ozone. Ozone depletion in the stratosphere is likely to be greater at high latitudes in winter than at lower latitudes in other seasons.[55] These variations in ozone distribution could create significant negative feedbacks affecting greenhouse warming.[56]

The principal method used to convert future concentration levels to estimates of future global warming is based on a one-dimensional radiative/convective model of the Earth-atmosphere system. As a consequence, the model cannot adequately simulate the effects of geographic or altitudinal variation in the distribution of these gases. Applying this model to the greenhouse problem requires scaling the total direct warming due to an increase in the concentration of each greenhouse gas to the levels estimated in each scenario. But the logarithmic scaling of CO_2 effects, the square root scaling of N_2O and CH_4, and the linear scaling of temperature response for other trace gases may not be completely appropriate over the entire range of concentrations considered in these scenarios. Applying fixed feedback factors to all scenarios may somewhat overstate these effects in extreme scenarios. Once the polar sea ice has melted in a combined with reductions in carbon monoxide emission from combustion. The changes in energy policy assumed in the Slow Build-up Scenario are likely to cause smaller but still significant reductions in the future warming commitment due to atmospheric build-up of nitrous oxide and tropospheric ozone as well.

In these scenarios, economic growth is not sacrificed to limit greenhouse gas emissions. A high-efficiency, low-emissions scenario such as the Slow Build-up

---

The longer the delay before preventive policies are identified, agreed upon, and implemented, the more extreme the policies imposed to stay within prudent bounds will have to be.

Scenario could give societies an additional thirty to sixty years to adapt to the unavoidable aspects of any given level of climate change and to find new technologies and other means to protect Earth's atmosphere from even more far-reaching changes.

Outcomes similar to the Slow Build-up Scenario described here have been derived in other studies using quite different methods and assumptions. The high-efficiency future illustrated by the Slow Build-up Scenario resembles that suggested by Goldemberg *et al.*[57] and by Rose *et al.*[58] While none of these scenarios eliminates the risk of significant future warming, they do offer the prospect of preventing the most extreme environmental and economic consequences of a greenhouse gas build-up.

The principal conclusion of *A Matter of Degrees* differs from that of the pathbreaking work by Seidel and Keyes,[59] who concluded in 1983 that the onset of global warming could not be significantly delayed by policies implemented in the near future. The current study reaches the opposite conclusion for several reasons. First, the scenarios tested here incorporated the effects of a wider range of assumptions about the possible rates of improvement in energy efficiency, about the level of biotic emissions of CO_2 and about the use of energy supply technologies that are alternatives to fossil-fuel combustion. Second, this study incorporates a range of assumptions about possible controls over the emissions of non-CO_2 greenhouse gases, while those tested by

Seidel and Keyes applied only to CO_2 emissions. Controlling the emissions of these other gases can help delay the onset of global warming significantly. Third, the study by Seidel and Keyes used a version of the Lacis equation to evaluate the warming effects of greenhouse gas build-up rather than the Ramanathan model used in this study. (Although the two models consider essentially the same greenhouse gases, they place different relative weights on build-up of each gas.) Ramanathan's model is used here because it employs a more recent and more appropriate set of absorption spectra.

The conclusions of *A Matter of Degrees* also have very different implications from those of the National Academy of Science study, *Changing Climate*, released in 1983.[60] That study, which focused principally on the warming effects of CO_2 build-up, suggested that nothing need be done in the near term and that a "wait and see" approach to climate change would be prudent public policy for now. The calculations reported in this study suggest exactly the opposite—that unless policies are implemented soon to limit greenhouse gas emissions, intolerable levels of global warming will result. As a consequence, the study of policy options for controlling the emissions of greenhouse gases must begin immediately and the choice of policies implemented in the next few decades could substantially affect the timing and magnitude of future global warming.

This study shows that policies affecting future patterns of energy use and CO_2 emissions will have the most effect on global warming during the next fifty years. This conclusion is partly a function of the larger role of CO_2 in the total warming and partly a product of the particular policies analyzed in this study. To determine its robustness, a broader range of policy strategies—in particular, more stringent policies limiting the production and use of CFCs—must be tested.

Further exercises with this model should also test the impacts on CO_2 emissions of a broader range of assumptions about the future biotic contribution to annual emissions. The impacts of changes in policies or in patterns of energy end-use in developing countries should be further disaggregated. The potential benefits of limiting methane, nitrous oxide, and tropospheric ozone should be examined using more realistic models of their future emissions, as these become available.

The limitations and uncertainties in the models notwithstanding, the conclusion that policy choices made today and implemented in the next several decades could radically slow impending climate change appears solid. The challenge now facing policy-makers and analysts interested in global warming is to go beyond the rough investigations reported here to identify country-specific or regional policy options that minimize the rates of future greenhouse gas emissions while sustaining high rates of economic growth. This analysis suggests that such policies exist and that greenhouse gas build-up can be controlled. If identified quickly, analyzed thoroughly, and optimized for the range of local conditions and national needs, such policies could allow a smooth transition to a high economic growth/low energy growth future.

Human societies have no historical or cultural records to help them deal with climate changes of the magnitude envisioned in this study. Only if the most vulnerable economic sectors and geographic regions are given time to prepare for those climatic changes that are unavoidable can societies hope to prosper under the rapidly changing climatic regimes. Fortunately, strong policies implemented in the next few decades could buy that time.

Postscript

Recent intergovernmental negotiations to develop a protocol to implement the Vienna Convention to Protect the Ozone Layer have focussed on alternative strategies for regulating chlorofluorocarbons (CFCs) to reduce the risk of stratospheric ozone depletion. As concern about the linked problems of ozone depletion and global warming increases, governments are becoming more willing to consider strong measures to limit the production and use of the most dangerous CFCs. Several approaches now under discussion are even more stringent than the proposals contained in the Slow Build-Up Scenario of **A Matter of Degrees.**

The U.S., Canadian, Swedish, Norwegian, and Finnish governments called in December 1986 for a freeze on all fully-halogenated CFCs and a phase-out of the consumption of these compounds to 95 percent below 1986 levels within about fifteen years. This proposal would cover production of CFC-11, CFC-12, CFC-113, and Halons 1211 and 1301. In March 1987, the European Economic Community proposed a freeze on production of CFC-11 and CFC-12 followed by a 20 percent reduction below 1986 production levels within six years. Adopting either proposal, or some compromise between them, will slow the build-up of CFCs even more than the measures simulated in this study. If such measures are combined with the aggressive policies outlined in the Slow Build-Up Scenario, the commitment to global warming of 1.5° to 4.5°C can be delayed even more than 60 years.

Dr. Irving M. Mintzer is Senior Associate and Director of WRI's Energy and CO$_2$ Projects. Formerly, he was an Associate Research Specialist in the Energy and Resources Group at the University of California at Berkeley and a Consultant to the International Institute for Applied Systems Analysis in Austria.

Appendix

Figure A-1. CO_2 Emissions From Fossil Fuel Use (Gigatons of Carbon per Year)

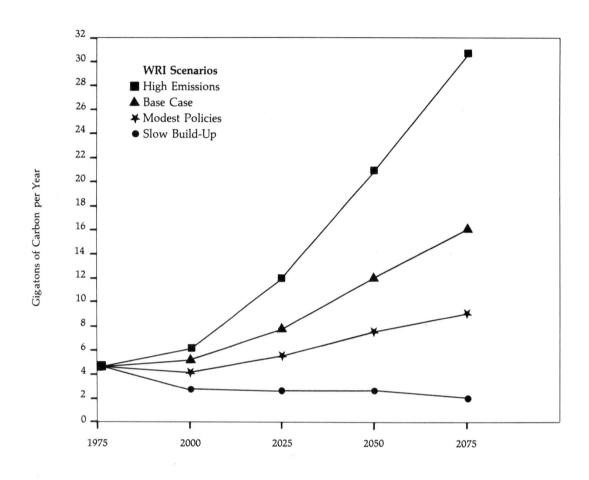

Figure A-2. CO_2 Emissions From Biotic Sources (Gigatons of Carbon per Year)

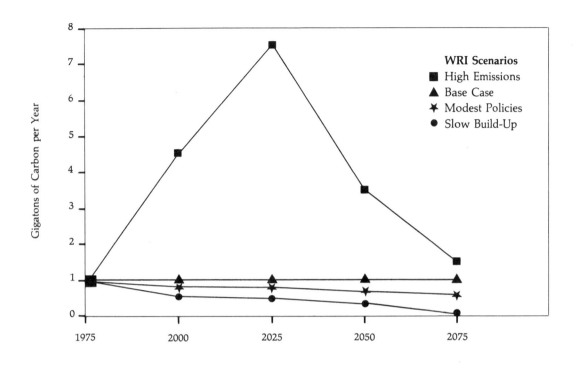

Figure A-3. Primary and Secondary Energy Supply in the Base Case Scenario (Exajoules Per Year)

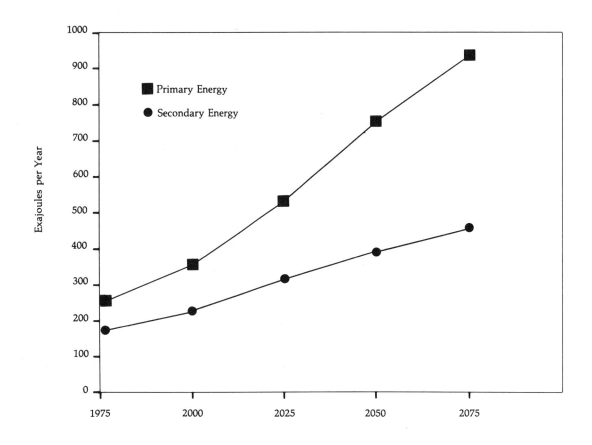

Table A-1   CO$_2$ Emissions from Fossil
Fuel Use
(Millions of Tons of
Carbon per Exajoule)

	Extraction & Production	Combustion	Total Emissions
Conventional Gas		13.8	13.8
Conventional Oil		19.7	19.7
Coal		26.9	26.9
Synthetic Oil	18.9	19.7	38.6
Synthetic Gas	26.9	13.8	40.7
Shale Oil	27.9	19.7	47.6

Source: J. Edmonds and J. Reilly, 1983b, ''Global Energy and CO$_2$ to the year 2050'' in *Energy Journal,* volume 4, no. 3, pp. 21–47.

Table A-2   Percentage of Total World Production of CFC-11 and CFC-12 in 1980

Region	CFC-11		
	Aerosol	Non-Aerosol	Total
Western Market Economies			
United States (1)	3.44%	34.51%	22.76%
Western Europe and Canada (2)	67.84%	37.57%	49.02%
Japan, Australia, and New Zealand (2)	16.96%	22.10%	20.16%
Sub Total (3)	88.25%	94.18%	91.94%
East Bloc Countries			
Soviet Union (4)	7.81%	3.91%	5.38%
Centrally Planned Europe	1.17%	0.59%	0.81%
Sub Total	8.98%	4.49%	6.19%
Developing Countries (5)			
Latin America	1.51%	0.71%	1.02%
Africa	1.18%	0.56%	0.79%
Other	0.08%	0.05%	0.06%
Sub Total	2.77%	1.33%	1.87%
World Total	100.00%	100.00%	100.00%

Region	CFC-12		
	Aerosol	Non-Aersol	Total
Western Market Economies			
United States (1)	2.85%	50.53%	30.15%
Western Europe and Canada (2)	55.93%	16.45%	33.33%
Japan, Australia, and New Zealand (2)	14.02%	16.49%	15.43%
Sub Total (3)	72.80%	83.47%	78.91%
East Bloc Countries			
Soviet Union (4)	20.72%	12.63%	16.09%
Centrally Planned Europe	3.11%	1.89%	2.41%
Sub Total	23.83%	14.52%	18.50%
Developing Countries (5)			
Latin America	1.85%	1.10%	1.42%
Africa	1.121%	0.71%	0.52%
Other	0.32%	0.20%	0.25%
Sub Total	3.37%	2.01%	2.59%
World Total	100.00%	100.00%	100.00%

Notes:
1. U.S. data are from the Annual Report of the U.S. International Trade Commission based on the regional distributions given in (RAND, 1986)
2. Data for W. Europe and Canada computed using regional distributions given in (RAND, 1986), Table 3.1, and the Total Reporting Companies Production given in (CMA, 1984). Production allocated by RAND to the category "Other Developed" is allocated here two-thirds to W. Europe and one-third to the Pacific region.
3. Total production data for each application are taken from Schedules 5 and 6 of (CMA, 1984).
4. Eastern European production is assumed to be 15% of Soviet production in each category. Soviet data for 1980 estimated from 1976 data reported by CMA, assuming continuation of 18% annual rate through 1980.
5. Developing country production in each region is calculated assuming the same fraction of total world production estimated by (RAND, 1986) in Table 3.1

Table A-3   Estimated Warming Commitments in 2075 Using Lacis vs. Ramanathan Approaches

Scenario	Lacis Approach	Ramanathan Approach
High Emissions	5.1-15.3°C	5.3-16.0°C
Base Case	3.3-9.9	2.9-8.6
Modest Policies	2.8-8.4	2.3-7.0
Slow Build-Up	2.0-6.0	1.4-4.2

Table A-4   Growth in Population and Labor Productivity in WRI Scenarios

Regional Growth Rates in Labor Productivity

Region	Approximate Annual Rate
USA	1.2%
Canada & W. Europe	1.6%
OECD Pacific	2.3%
USSR/E. Europe	1.3%
China/*et al.*	1.9%
Middle East	1.9%
Africa	1.6%
Latin America	1.9%
S. & E. Asia	1.8%
Industrial Countries	1.2-2.3%
Developing Countries	1.6-1.9%

Global Population
Millions of Persons

Region	1975	2000	2025	2050	2075
USA	214	260	290	290	290
Canada & W. Europe	405	480	520	540	540
OECD Europe	128	150	160	150	150
USSR/E. Europe	395	430	470	500	510
China/*et al.*	911	1300	1600	1700	1700
Middle East	81	170	280	360	410
Africa	399	890	1600	2200	2700
Latin America	313	520	720	850	900
S. & E. Asia	1130	1900	2600	3100	3400
All Regions	3980	- 6100	8240	9690	10600

Population Growth Rates	1975–2000	1975–2025	1975–2050	1975–2075
Global	1.7%	1.5%	1.2%	1.0%
Developing Country				1.2%
Industrial Country				0.3%

Table A-5 Future Energy Prices in WRI Scenarios
(All Prices in Constant 1975 US$ Per Gigajoule)

Scenario	Oil			Gas			Solids		
	1975 Price	2075 Price	Growth Rate	1975 Price	2075 Price	Growth Rate	1975 Price	2075 Price	Growth Rate
High Emissions	$1.84	$4.68	0.9%	$0.63	$3.21	1.6%	$0.51	$1.04	0.7%
Base Case	$1.84	$5.03	1.0%	$0.63	$3.06	1.6%	$0.51	$0.88	0.5%
Modest Policies	$1.84	$5.44	1.1%	$0.63	$2.71	1.5%	$0.51	$0.78	0.4%
Slow Build-Up	$1.84	$5.63	1.1%	$0.63	$1.78	1.0%	$0.51	$1.16	0.8%

Table A-6 Annual Energy Use Per Capita in WRI Scenarios
(Energy Use in Gigajoules; 1975–2025 and 1975–2075 Growth Rates in Parentheses)

Scenario	Industrialized Countries			Developing Countries		
	1975	2025	2075	1975	2025	2075
High Emissions	160	290 (1.2%)	519 (1.2%)	18	43 (1.8%)	92 (1.6%)
Base Case	160	220 (0.64%)	310 (0.66%)	18	30 (1.0%)	52 (1.1%)
Modest Policies	160	180 (0.24%)	210 (0.27%)	18	24 (0.58%)	35 (0.67%)
Slow Build-Up	160	120 (−0.57%)	110 (−0.37%)	18	12 (−0.81%)	11 (−0.49%)

Table A-7
Scenarios of Energy Supply by Fuel Type
(Energy Supply Given in Exajoules)

	Oil	Gas	Coal	Nuclear	Biomass	Solar	Hydro	Total
Energy Supply Scenarios: 1975								
Historical Data	119	41	76	7	0	0	9	252
Energy Supply Scenarios: 2000								
Scenario	**Oil**	**Gas**	**Coal**	**Nuclear**	**Biomass**	**Solar**	**Hydro**	**Total**
High Emissions	78	66	168	13	12	0	57	390
Base Case	82	63	115	11	13	0	54	340
Modest Policies	80	61	88	11	16	0	54	310
Slow Build-Up	68	58	33	9	27	0.2	41	240
Energy Supply Scenarios: 2025								
Scenario	**Oil**	**Gas**	**Coal**	**Nuclear**	**Biomass**	**Solar**	**Hydro**	**Total**
High Emissions	93	67	393	32	21	2	104	710
Base Case	68	59	243	22	22	13	91	520
Modest Policies	78	55	146	19	23	13	90	420
Slow Build-Up	74	53	19	10	36	15	40	250
Energy Supply Scenarios: 2050								
Scenario	**Oil**	**Gas**	**Coal**	**Nuclear**	**Biomass**	**Solar**	**Hydro**	**Total**
High Emissions	113	59	762	42	33	26	118	1150
Base Case	56	57	429	26	33	27	118	750
Modest Policies	67	55	233	21	31	23	118	550
Slow Build-Up	74	62	16	8	45	18	56	280
Energy Supply Scenarios: 2075								
Scenario	**Oil**	**Gas**	**Coal**	**Nuclear**	**Biomass**	**Solar**	**Hydro**	**Total**
High Emissions	70	53	1196	78	52	46	120	1610
Base Case	30	51	620	38	44	38	119	940
Modest Policies	36	56	322	28	39	29	119	630
Slow Build-Up	48	57	18	10	55	20	56	260

Notes:
1. Primary energy supplied from synthetic oil and synthetic gas is counted in the energy supply from coal. Primary energy supplied from shale oil is counted with oil.
2. Nuclear, solar, and hydro energy supplies are counted at their fuel-equivalent value.
3. Biomass refers only to annually cycled, commercial biomass supplies.

Table A-8   Comparison of High Emissions Scenario with Other
High Case Projections of Future Energy Demand, 2025

| Source | Population in billions | Global Primary Energy Demand EJ Per Year | Growth Rates | |
			Pop.	GDP Per Capita
World Energy Conf. (1978)	9	1060	1.8%	2.0%
IIASA (1981)	8	1100	1.2%	2.1%
Seidel & Keyes (1963)	7.4	[2]800	1.2%	1.7%
Edmonds & Reilly (1984)	7.4	[2]1100	1.2%	2.9%
High Emissions Case (This Study)	8.2	710	1.5%	1.4%

Sources:
J. Edmonds et al., 1984, ''Case A,'' in ''An Analysis of Possible Future Atmospheric Retention of Fossil Fuel CO_2'' Report No. DOE/OR/21400-1, Washington, D.C.
S. Seidel and D. Keyes, 1983, ''High Fossil Scenario.''
IIASA, 1981, ''High Scenario'' in *Energy in a Finite World,* W. Hafele, ed., Ballinger, Cambridge, MA.
World Energy Conference, 1978, ''High Growth Scenario'' in *World Energy Resources 1985–2020,* IPC Science and Technology Press, Guildford, U.K.

Table A-9   Comparison of Modest Policies and Controlled Risk Scenarios
with Other Low Growth Projections of Future Energy Demands, 2025

Source	Population in billions	Global Primary Energy Demand EJ Per Year	Growth Rates Pop.	Growth Rates GDP Per Capita
World Energy Conf. (1978)	9	840	1.8%	1.2%
IIASA (1981)	8	710	1.2%	1.1%
Colombo and Bernardi (1979)	8.0	470	1.2%	1.2%
Lovins, *et al.* (1981)	8.0	170	1.2%	1.1%
Seidel & Keyes (1983)	7.4	[2]700	1.2%	1.7%
Edmonds & Reilly (1984)	7.4	540	1.2%	2.9%
Rose, *et al.* (1984)	[2]7.4	[2]400	1.2%	*
Goldemberg, *et al.* (1985)	[2]7	[2]360	1.1%	*
MODEST POLICIES (This Study)	8.2	420	1.5%	1.4%
SLOW BUILD-UP (This Study)	8.2	250	1.5%	1.4%

Sources:
J. Emonds *et al.*, 1984, ''Case C''
Colombo, U. and O. Bernardini, 1979 in *A Low Energy Growth Scenario and the Prospectives for W. Europe,* C.E.C., Brussels.
Lovins, *et al.,* 1981. *Least Cost Energy; Solving the CO_2 Problem,* Brick House Press, Andover, MA.
J. Goldemberg *et al.,* 1985 (Data for 2020)
D. Rose, *et al.,* 1983, ''Case J''
IIASA, 1981, ''Low Scenario''
S. Seidel and D. Keyes, 1983, ''High Renewable Scenario''
World Energy Conference, 1978, ''Low Growth Scenario''

Table A-10   Annual Rate of Growth in Coal Use

Scenario	1975–2000	2000–2025	2025–2050	2050–2075	1975–2075
High Emissions	3.3%	3.5%	2.7%	1.8%	2.8%
Base Case	1.7%	3.0%	2.3%	1.5%	2.1%
Modest Policies	0.6%	2.0%	1.9%	1.3%	1.5%
Slow Build-Up	−4.8%	−0.6%	−0.7%	0.5%	−1.4%

Table A-11 Projected Trace Gas Concentrations in the Base Case Scenario

Year	CO_2 ppmv	N_2O ppbv	CH_4 ppbv	Ozone % change	CFC11 ppbv	CFC12 ppbv
1980	339	301	1570	0	0.170	0.285
1990	353	307	1734	3	0.303	0.521
2000	370	313	1916	5	0.471	0.793
2010	388	322	2116	8	0.648	1.080
2020	409	333	2338	10	0.826	1.377
2030	433	349	2582	13	1.005	1.687
2040	461	368	2852	15	1.190	2.014
2050	494	392	3151	15	1.379	2.359
2060	531	420	3480	15	1.573	2.720
2075	595	470	4040	15	1.873	3.291

Table A-12 Projected Trace Gas Concentrations in the High Emissions Scenario

Year	CO_2 ppmv	N_2O ppbv	CH_4 ppbv	Ozone % change	CFC11 ppbv	CFC12 ppbv
1980	339	301	1570	0	0.170	0.285
1990	355	308	1778	3	0.303	0.521
2000	381	317	2167	5	0.512	0.827
2010	414	329	2774	8	0.815	1.225
2020	456	347	3551	10	1.196	1.714
2030	508	372	4545	13	1.651	2.295
2040	564	404	5818	15	2.198	2.983
2050	627	446	7448	15	2.897	3.828
2060	697	496	9534	15	3.859	4.904
2075	817	590	13808	15	5.749	6.906

Table A-13   Projected Trace Gas Concentrations in the Modest Policies Scenario

Year	CO_2 ppmv	N_2O ppbv	CH_4 ppbv	Ozone % change	CFC11 ppbv	CFC12 ppbv
1980	339	301	1570	0	0.170	0.285
1990	353	306	1734	3	0.302	0.517
2000	368	312	1892	5	0.464	0.748
2010	383	317	2039	8	0.635	0.987
2020	399	325	2197	10	0.796	1.232
2030	417	334	2368	13	0.959	1.488
2040	436	344	2551	15	1.127	1.756
2050	457	357	2749	15	1.299	2.039
2060	480	372	2962	15	1.474	2.333

Table A-14   Projected Trace Gas Concentrations in the Slow Build-up Scenario

Year	CO_2 ppmv	N_2O ppbv	CH_4 ppbv	Ozone % change	CFC11 ppbv	CFC12 ppbv
1980	339	301	1570	0	0.170	0.285
1990	352	304	1692	3	0.282	0.500
2000	362	305	1778	5	0.381	0.693
2010	371	306	1869	8	0.470	0.875
2020	379	306	1965	10	0.547	1.046
2030	387	306	2065	13	0.615	1.206
2040	395	307	2153	15	0.674	1.355
2050	403	306	2225	15	0.726	1.494
2060	410	306	2299	15	0.771	1.625
2075	419	307	2416	15	0.829	1.805

Notes

1. Arhennius, S. 1896. "On the influence of Carbonic acid in the air upon the temperature of the ground," in Phil. Mag., 41, 237.

2. See, for example, Goldemberg, et al., 1985, "An End Use Oriented Global Energy Strategy," in Annual Review of Energy, Palo Alto, California, pp. 613–688; and Rose et al., 1983, Global Energy Futures and CO$_2$-Induced Climate Change. MITEL 83-015, MIT Energy Lab, Cambridge, MA.

3. Descriptions of these models have been published separately and widely peer-reviewed. Their limitations and weaknesses are discussed in Section VI.

   Cicerone R. and R. Dickinson, 1986. "Future Global Warming from Atmospheric Trace Gases," in Nature, volume 319, January 9, 1986, pp. 109–115.

   Edmonds, J. and J. Reilly, 1983a. "A Long-Term Global Energy-Economic Model of Carbon Dioxide Release from Fossil Fuel Use," in Energy Economics, volume 5, number 2, pp. 74–88.

   Edmonds, J., and J. Reilly, 1983b. "Global Energy and CO$_2$ to the Year 2050," in Energy Journal, volume 4, number 3, pp. 21–47.

   Edmonds, J., J. Reilly, R. Gardner, and A. Brenkert, 1985. Uncertainty in Future Global Energy Use and Fossil Fuel CO$_2$ Emissions 1975 to 2075, Institute for Energy Analysis, Oak Ridge Associated Universities, Washington, D.C.

   ICF Incorporated, 1986. "Relationships Among CFC Use, CFC Emissions, and Banked CFCs," Annex C in Scenarios of CFC Use: 1985–2075, ICF Incorporated, Washington, D.C.

   Quinn et al., 1986. Projected Use, Emissions, and Banks of Potential Ozone-Depleting Substances, RAND Note number N-2282-EPA, Rand Corporation, Santa Monica, CA.

   Ramanathan et al., 1985. "Trace Gas Trends and their Potential Role in Climate Change," in Journal of Geophysical Research, volume 90, number D3, pp. 5547–5566.

   Weiss, R.F., 1981. "The Temporal and Spatial Distribution of Tropospheric Nitrous Oxide," in Journal of Geophysical Research, volume 86, number C8, August 20, 1981, pp. 7185–7195.

4. MacDonald, G., 1986. Climate Change and Acid Rain, Mitre Corporation, Maclean, VA.

   Seidel, S. and D. Keyes, 1983. Can We Delay a Greenhouse Warming? U.S. Environmental Protection Agency, Washington, D.C.

   WMO, 1986a. Atmospheric Ozone 1985: Assessment of Our Knowledge of the Processes Controlling Its Present Distribution and Change, WMO Report No. 16, Geneva, 1986.

   WMO, 1986b. Report of the International Conference on the Assessment of the Role of Carbon Dioxide and of Other Greenhouse Gases in Climate Variations and Associated Impacts, A conference sponsored by the United Nations Environment Programme, the International Council of Scientific Unions, and the World Meteorological Organization. Report published by the World Climate Programme, World Meteorological Organization, Geneva, Switzerland.

5. J. Hansen, A. Lacis, D. Rind, and D. Russell, 1984. "Climate Sensitivity to Increasing Greenhouse Gases," in J. Titus and M. Barth, ed., Sea Level Rise to the Year 2100, Van Nostrand and Co., New York.

6. Clark, W. et al., 1982. "The Carbon Dioxide Question: A Perspective for 1982," in W. Clark, ed., Carbon Dioxide Review: 1982, Oxford University Press, New York, pp. 3–43.

7. Ramanathan et al., 1985. The authors estimate the direct radiative effect of emissions from 1860 to 1980 to be approximately 0.8°C. The range quoted in the text is an estimate of the equilibrium warming, after consideration of feedback effects.

8. NAS, 1979. *Carbon Dioxide and Climate: A Scientific Assessment*, National Research Council, Washington, D.C.

NAS, 1983. *Changing Climate*, the Report of the Carbon Dioxide Assessment Committee, U.S. National Academy of Sciences, National Academy Press, Washington, D.C.

9. U.S. Department of Energy, 1986. ''Human Alterations of the Global Carbon Cycle and the Projected Future,'' in *Atmospheric Carbon Dioxide and the Global Carbon Cycle*, J. Trabalka, ed., Report No. DOE/ER-0239, National Technical Information Service, Springfield, VA, December 1985.

Dickinson, R., 1986. ''How will Climate Change?,'' in Bolin, Doos, Warrick, and Jager, eds., *The Greenhouse Effect, Climate Change and Ecosystems*, John Wiley and Sons, Chichester and New York, pp. 207–270.

10. WMO, 1986b.

11. Emissions are considered ''prompt'' if they occur within twelve months of the date of manufacture of these chemicals. ''Banked'' emissions refer to CFCs embodied in long-lived products (such as refrigerators and air conditioners) whose chemical contents are not released to the atmosphere until the product is disposed of. These delayed or ''banked'' releases may not take place for twenty to one hundred years after the chemicals are manufactured.

12. The climate systems was *not* in equilibrium in 1980. A substantial fraction of the warming commitment due to greenhouse gas build-up since 1860 has not yet been realized.

13. Edmonds, J. and J. Reilly, 1983a,b. 1986.

Applications of this model include Seidel and Keyes (1983), Rose *et al* (1983), and Edmonds, J. and J. Reilly 1985, ''Future Global Energy and Carbon Dioxide Emissions'' in *Atmospheric Carbon Dioxide and the Global Carbon Cycle*, J. Trabalka, ed., Report No. DOE/ER-039, National Technical Information Service, Springfield, VA, December 1985.

14. In the Edmonds-Reilly model, changes in labor productivity over time (measured in dollars of GNP person-year of labor) reflect an assumption of increasing efficiency of labor as an input to production. This assumed increase is believed to reflect a combination of factors, including higher levels of education, better diet and health care, etc. No measure of the costs of these improved living conditions is incorporated in the model.

15. The end-use efficiency parameter reflects the annual rate at which improvements in technology reduce the amount of energy required per unit of output in each end-use sector.

16. The energy supply efficiency factor reflects the annual rate (given in percent per year) at which improvements in technology (for instance oil-drilling equipment) reduce the cost of supplying a unit of fuel to the point of end-use (given in constant dollars per gigajoule). For non-renewable energy resources, the minimum extraction cost of fuel increases over time as lower-quality supplies are exploited. The supply efficiency factor tends to decrease the cost of energy over time while increasing resource scarcity tends to increase supply costs.

17. Clark, *et al.*, 1982.

18. WMO, 1986a.

19. Weiss, R. and H. Craig, 1976. ''Production of Atmospheric Nitrous Oxide by Combustion,'' Geophysical Research letters, volume 3, pp. 751–753.

20. Considerable uncertainty exists about the contribution to anthropogenic N_2O emissions due to the use of nitrogenous fertilizers and to deforestation. Use of nitrogenous fertilizers is likely to grow as a global population increases. Emissions of N_2O from the decomposition of fertilizers may be a significant source of N_2O in the future.

21. Bowden, W.B. and F.H. Bormann, 1986. ''Transport and Loss of Nitrous Oxide in Soil Water After Forest Clear-Cutting,'' *Science*, Vol. 233, No. 4766, pp. 867–869.

22. Craig, H. *et al.*, 1976. Paper presented at the Symposium on the Terrestrial Nitrogen Cycle and Possible Atmospheric Effects, American Geophysical Union, Washington, D.C.

23. Weiss, R.F., 1981. ''The Temporal and Spatial Distribution of Tropospheric Nitrous Oxide,'' in *Journal of Geophysical Research*, volume 86, number C8, August 20, 1981, pp. 7185–7195.

24. Weiss, 1981.

25. Quinn *et al.*, 1986.

26. ICF Incorporated, 1986.

27. Cicerone R. and R. Dickinson, 1986.

28. See, e.g., WMO, 1986a.

29. This cooling of the stratosphere due to a greenhouse warming in the troposphere may reduce the risk of ozone depletion by slowing somewhat the rate of certain key reactions which are believed to be involved in stratospheric processes leading to the catalytic destruction of ozone.

30. WMO, 1986b.

31. Ramanathan *et al.*, 1985.

32. See for example, NAS, 1983; and WMO, 1986.

33. Lacis *et al.*, 1981.

34. The Lacis equation used to calculate the radiative forcing is as follows:

$$\begin{aligned}
Teq\,(^\circ C) = {} & 0.57*(CH_4(t)^{0.5}) + 2.8*(N_2O(t)^{0.6}) \\
& -0.057*(CH_4(t)*N_2O(t)) \\
& + 0.15 * CFC11(t) \\
& +0.18 * CFC12(t) \\
& + 2.5* \ln\,[1 + 0.005\,(CO_2(t) - 300) \\
& + 0.0001*\,(CO_2(t) - 300)\,]
\end{aligned}$$

35. Motivated by concerns other than global warming, a number of studies have tried recently to project the likely future evolution of the global energy section (e.g., World Energy Conference, 1978; IIASA, 1981; Nordhaus and Yohe, 1983; Goldemberg, *et al.* 1985).

36. In the Edmonds-Reilly model, the price of fuels is computed as a sum of extraction or production costs, environmental costs, and transport costs. The relative prices of fuels determine their market share in each region. In the scenarios analyzed in this study, extraction and transport costs remain constant but environmental costs vary substantially. All prices in this report are given in constant 1975 US$.

37. Energy from nuclear, solar, and hydro resources are counted at their fuel-equivalent value.

38. World Energy Congress, 1983. *Energy 2000–2020: World Prospects and Regional Stresses.* J. R. Frisch, ed., World Energy Congress, London, England.

39. The Edmonds-Reilly model includes fission electric but not fusion electric systems (whose future costs are difficult to estimate) as a component of future electricity supplies.

40. For a more detailed discussion of this multilateral investment plan, see WRI, *Tropical Forests: A Call for Action*, World Resources Institute, Washington, D.C., 1985.

41. See e.g., Goldemberg J., *et al.*, 1985. ''An End Use Oriented Global Energy Strategy,'' in *Annual Review of Energy*, volume 10, Annual Review Press, Palo Alto, CA, pp. 613–688.

42. Ausubel, J. and W. Nordhaus, 1983. ''A Review of Estimates of Future CO_2 Emissions,'' in NAS, 1983.

43. Goodman, G., 1983. ''Societal Problems in Meeting Current and Future Energy Needs,'' in *Ambio*, volume 12, number 2, pp. 97–101.

44. Williams, R. *et al.*, 1984. *Overview of An End-Use Oriented Global Energy Strategy*, A paper presented at the Symposium on Greenhouse Problem Policy Options, Hubert Humphrey Institute of Public Affairs, Minneapolis, MN, May 29–31, 1984.

45. Manne, A.and L. Schratenholzer, 1984. ''International Energy Workshop: A Summary of the 1983 Poll Responses,'' in *Energy Journal*, volume 5, number 1, pp. 45–65.

46. Laurmann, J. 1984. ''Market Penetration as an Impediment to Replacement of Fossil Fuel in the CO_2 Environmental Problem,'' Paper presented at the Symposium on Greenhouse Problem Policy Options, Hubert Humphrey Institute of Public Affairs, Minneapolis, MN, May 29–31, 1984.

47. Williams *et al.*, 1987. *Energy for a Sustainable World.* World Resources Institute, Washington, D.C.

48. Jaeger, J. W. 1984. ''Impacts on the Energy Sector,'' in H. Meinl, ed., *Socioeconomic Impacts of Climatic Changes Due to a Doubling of Atmospheric CO_2 Content*, EEC, Brussels, Belgium, pp. 396–418.

49. Nader, L. *et al.*, 1979. *Impacts on Consumption, Location, and Occupational Patterns on Energy Demand: A Report to the Demand Panel of the U.S. National Academy of Sciences Committee on Nuclear and Alternative Energy Systems*, National Research Council, Washington, D.C.

50. Edmonds, J. and J. Reilly, 1985.

51. U.S. Department of Energy, 1986.

52. Keepin, W., I. Mintzer, and Kristoferson, 1986. ''Emissions of CO_2 into the Atmosphere,'' in B. Bolin *et al.*, *The Greenhouse Effect, Climatic Change, and Ecosystems: A Synthesis of the Present Knowledge*, Wiley and Sons, Chichester, U.K., in press.

53. Thompson, A. and R. Cicerone, 1985. ''Atmospheric CH_4, CO, and OH from 1860 to 1985,'' submitted to *Nature*, in press.

54. Keepin, Mintzer, and Kristoferson, 1986.

55. WMO, 1986b.

56. Wang and Sze, private communication, 1986.

57. Goldemberg *et al.*, 1985.

58. Rose *et al.*, 1983.

59. Seidel, S. and D. Keyes, 1983.

60. National Academy of Sciences, 1983.

WRI PUBLICATIONS ORDER FORM

ORDER NO.	TITLE	QTY	TOTAL $
S710	*A Matter of Degrees: The Potential for Controlling the Greenhouse Effect* by Irving M. Mintzer, 1987, $10.00		
S792	*Skimming the Water: Rent-Seeking and the Performance of Public Irrigation Systems* by Robert Repetto, 1986, $10.00		
S783	*The Sky is the Limit: Strategies for Protecting the Ozone Layer* by Alan S. Miller and Irving M. Mintzer, 1986, $10.00		
S781	*Double Dividends? U.S. Biotechnology and Third World Development* by John Elkington, 1986, $10.00		
B719	*Bordering on Trouble: Resources and Politics in Latin America* edited by Andrew Maguire and Janet Welsh Brown, 1986, $14.95 (paperback)		
S784	*Troubled Waters: New Policies for Managing Water in the American West* by Mohamed T. El-Ashry and Diana C. Gibbons, 1986, $10.00.		
S712	*Growing Power: Bioenergy for Development and Industry* by Alan S. Miller, Irving M. Mintzer, and Sara H. Hoagland, 1986, $10.00.		
S725	*Down to Business: Multinational Corporations, the Environment, and Development* by Charles S. Pearson, 1985, $10.00		
B723	*The Global Possible: Resources, Development, and the New Century* edited by Robert Repetto, 1986, $13.95 (paperback); $45.00 (cloth)		
B732	*World Enough and Time: Successful Strategies for Resource Management* by Robert Repetto, 1986, $5.95 (paperback); $16.00 (cloth)		
S724	*Getting Tough: Public Policy and the Management of Pesticide Resistance* by Michael Dover and Brian Croft, 1984, $10.00		
S714	*Field Duty: U.S. Farmworkers and Pesticide Safety* by Robert F. Wasserstrom and Richard Wiles, 1985, $10.00		
S716	*A Better Mousetrap: Improving Pest Management for Agriculture* by Michael J. Dover, 1985, $10.00		
S717	*The American West's Acid Rain Test* by Philip Roth, Charles Blanchard, John Harte, Harvey Michaels, and Mohamed El-Ashry, 1985, $10.00		
S715	*Paying the Price: Pesticide Subsidies in Developing Countries* by Robert Repetto, 1985, $10.00		
S776	*The World Bank and Agricultural Development: An Insider's View* by Montague Yudelman, 1985, $10.00		
S731	*Tropical Forests: A Call for Action*, 1985 by WRI, The World Bank and UNDP, $12.50		
S726	*Helping Developing Countries Help Themselves: Toward a Congressional Agenda for Improved Resource and Environmental Management in the Third World* (a WRI working paper) by Lee M. Talbot, 1985, $10.00		
B780	*World Resources 1987*, $16.95 (paperback); $32.95 (cloth)		
	WRI SUBSCRIPTION $50. ($70 for overseas)		
	TOTAL		

Name _____ (last) _____ (first) _____

Place of Work _____

Street Address _____

City/State _____ DEGREES87 _____ Postal Code/Country _____

Please send check or money order (U.S. dollars only) to WRI Publications, P.O. Box 620, Holmes, PA 19043-0620, U.S.A.

BECOME A WRI SUBSCRIBER

■ Receive all WRI Policy Studies, all WRI research reports, and occasional publications for calendar year 1987—including *World Resources 1987*. If you join in March, for example, you will immediately receive publications issued in January and February. $50.00. ($70.00 outside of the United States.)

■ Discounts available for bulk orders.

WRI PUBLICATIONS ORDER FORM

ORDER NO.	TITLE	QTY	TOTAL $
S710	*A Matter of Degrees: The Potential for Controlling the Greenhouse Effect* by Irving M. Mintzer, 1987, $10.00		
S792	*Skimming the Water: Rent-Seeking and the Performance of Public Irrigation Systems* by Robert Repetto, 1986, $10.00		
S783	*The Sky is the Limit: Strategies for Protecting the Ozone Layer* by Alan S. Miller and Irving M. Mintzer, 1986, $10.00		
S781	*Double Dividends? U.S. Biotechnology and Third World Development* by John Elkington, 1986, $10.00		
B719	*Bordering on Trouble: Resources and Politics in Latin America* edited by Andrew Maguire and Janet Welsh Brown, 1986, $14.95 (paperback)		
S784	*Troubled Waters: New Policies for Managing Water in the American West* by Mohamed T. El-Ashry and Diana C. Gibbons, 1986, $10.00.		
S712	*Growing Power: Bioenergy for Development and Industry* by Alan S. Miller, Irving M. Mintzer, and Sara H. Hoagland, 1986, $10.00.		
S725	*Down to Business: Multinational Corporations, the Environment, and Development* by Charles S. Pearson, 1985, $10.00		
B723	*The Global Possible: Resources, Development, and the New Century* edited by Robert Repetto, 1986, $13.95 (paperback); $45.00 (cloth)		
B732	*World Enough and Time: Successful Strategies for Resource Management* by Robert Repetto, 1986, $5.95 (paperback); $16.00 (cloth)		
S724	*Getting Tough: Public Policy and the Management of Pesticide Resistance* by Michael Dover and Brian Croft, 1984, $10.00		
S714	*Field Duty: U.S. Farmworkers and Pesticide Safety* by Robert F. Wasserstrom and Richard Wiles, 1985, $10.00		
S716	*A Better Mousetrap: Improving Pest Management for Agriculture* by Michael J. Dover, 1985, $10.00		
S717	*The American West's Acid Rain Test* by Philip Roth, Charles Blanchard, John Harte, Harvey Michaels, and Mohamed El-Ashry, 1985, $10.00		
S715	*Paying the Price: Pesticide Subsidies in Developing Countries* by Robert Repetto, 1985, $10.00		
S776	*The World Bank and Agricultural Development: An Insider's View* by Montague Yudelman, 1985, $10.00		
S731	*Tropical Forests: A Call for Action*, 1985 by WRI, The World Bank and UNDP, $12.50		
S726	*Helping Developing Countries Help Themselves: Toward a Congressional Agenda for Improved Resource and Environmental Management in the Third World* (a WRI working paper) by Lee M. Talbot, 1985, $10.00		
B780	*World Resources 1987*, $16.95 (paperback); $32.95 (cloth)		
	WRI SUBSCRIPTION $50. ($70 for overseas)		
	TOTAL		

Name   (last)   (first)

Place of Work

Street Address

City/State   **DEGREES87**   Postal Code/Country

Please send check or money order (U.S. dollars only) to WRI Publications, P.O. Box 620, Holmes, PA 19043-0620, U.S.A.

BECOME A WRI SUBSCRIBER

■ Receive all WRI Policy Studies, all WRI research reports, and occasional publications for calendar year 1987—including *World Resources 1987*. If you join in March, for example, you will immediately receive publications issued in January and February. $50.00. ($70.00 outside of the United States.)

■ Discounts available for bulk orders.